Lecture Notes in Energy 10

For further volumes:
http://www.springer.com/series/8874

Wei-Chiang Hong

Intelligent Energy Demand Forecasting

Springer

Wei-Chiang Hong
Department of Information Management
Oriental Institute of Technology
New Taipei City
Taiwan R.O.C.

ISSN 2195-1284 ISSN 2195-1292 (electronic)
ISBN 978-1-4471-4967-5 ISBN 978-1-4471-4968-2 (eBook)
DOI 10.1007/978-1-4471-4968-2
Springer London Heidelberg New York Dordrecht

Library of Congress Control Number: 2013933361

Printed on acid-free paper

Springer is part of Springer Science+Business Media (www.springer.com)

To my parents, who raised me, I was in need of them during my academic career.

To my wife, Su-Mei, with great love and respect.

To my son, Chang-Hsin Edward, for his constant understanding and kind encouragement.

To my beloved family and friends.

To everyone who has the underappreciated problem in academic career, please never give up yourself;
never lose hope for future.

Acknowledgements

Along with the rapid development of intelligent computing for multidisciplinary applications, there is a shortage of books focusing on energy demand forecasting. Many papers have been published in the literature, but no available book contains some critical intelligent computing technology for deeper discussions. The idea of writing this book came to my mind 2 or 3 years ago while I was working hard in publishing my research results to promote to worldwide academic society; however, the time was too limited to write such a book. In the late spring of 2011, Springer sent their energy publishing invitation; in the meanwhile, the National Science Council (NSC, Taiwan) also announced its academic book writing sponsor regulation. Thus, I had the very opportunity to apply the grant to finish my book. Indeed, my appreciation goes to Springer and NSC (sponsored grant number: NSC 101-2410-H-161-001).

I truly appreciate the friendship of my friends and colleagues among the faculty of the department of information management at Oriental Institute of Technology (Taiwan) for having created a pleasant working environment. In addition, many thanks to my friends among the editorial board of *Neurocomputing*, *Applied Soft Computing*, *Mathematical Problems in Engineering*, and *International Journal of Applied Evolutionary Computation* for their robust support and novel ideas. Special thanks go to my friends Professor Yucheng Dong (Sichuan University, China) and Professor Wen Yu Zhang (Zhejiang University of Finance and Economics, China) for their strong encouragement to live and stay in Taiwan, even if my significant contributions are underappreciated to be promoted to a high reputed university.

In addition, I'll be responsible for all details in this book; of course, I'd like to receive any constructive suggestion if existing any controversial issue, rather than unmeaning wording. While I listened attentively to the "Past Anecdotes" of France Royals from the short, fat tourist guide of Chateau de Versailles and the Louvre, the impression in my mind is not only his detail but also the responsibility for his job and pride in his country that he demonstrated I think this is the key successful factor for a great nation to remain competitive and become unshakeable as time goes by.

Finally, I owe my deepest appreciation to my parents, Mr. Hsien-Long Hong and Mrs. Hsu-Fang Chen, for their steady support and endless love. I'd like to express my sincere gratitude to my beloved wife Mrs. Su-Mei Tsai for her deepest love and taking care of our cute, clever son Chang-Hsin Edward, constant understanding, and being together with me in the whole research journey. I also would express my appreciation to my cute, clever son, Chang-Hsin Edward, for his kind encouragement and understanding my plight, busying to play the proper role among researching, teaching, and housekeeping. All of them have always been a source of inspiration. Without their support through these years, I would not be able to afford the luxury of pursuing my own research interests.

Contents

Chapter 1
Introduction

For an export-oriented economy or an energy-limited economy, like Taiwan, economic development mainly depends on the supply level of electric energy; particularly, most production activities for each industry also depend on its available level. As the economic development is proceeding vigorously, the electric energy demand in industries, in commerce, and in people's residential uses has also significantly increased. To ensure that electric energy is usable for all electricity users (i.e., meeting users' demands) will be an important challenge for the electric energy industry. The availability and reliability of electric energy become the most premier issue in energy policy making. Therefore, accurate electric load forecasting is quite an important guide for effective implementations/actions of energy policies. The policy makers desire to look for founded forecasts to well plan the new electric power facilities investments and to determine the import and export amounts.

In the meanwhile, along with the recent privatization and deregulation of the electricity industry, the reliance and accuracy of future electricity demand forecasting have received growing attention, particularly in the areas of electricity load planning, energy expenditure/cost economy, and secure operation fields, in regional and/or national systems. For electricity load reliance, electricity providers face increasing competition in the demand market and must pay increased attention to electricity quality, including unit commitment, hydrothermal coordination, short-term maintenance, interchange and transaction evaluation, network power flow dispatched optimization, and security strategies. On the other hand, inaccurate electricity load forecasting may increase operating costs [1–5]. Bunn and Farmer [4] point out that a 1 % increase in forecasting error implied a £10 million increase in operating costs. Hence, overestimation of future load results in unnecessary spinning reserve and, furthermore, is not accepted by international energy networks owing to excess supply. In contrast, underestimation of load causes failure in providing sufficient reserve and implies high costs in the peaking unit. Because buying at the last minute from other suppliers is expensive, it is necessary for international electricity production cooperation that every member is able to forecast its demands accurately. However, it is complex to predict the electric load, primarily due to the various influencing factors, such as climate factors, social activities, and seasonal factors [6]. Climate factors depend on the temperature and humidity; social

W.-C. Hong, *Intelligent Energy Demand Forecasting*, Lecture Notes in Energy 10,
DOI 10.1007/978-1-4471-4968-2_1, © Springer-Verlag London 2013

factors imply human social activities including work, school, and entertainment affecting the electric load; seasonal factors then include seasonal climate change and load growth year after year.

1.1 Traditional Approaches for Electric Load Forecasting

In the last few decades, there are widespread investigations with regard to the efforts proposed to improve the accuracy of electricity load forecasting. One such method is a weather insensitive approach that uses historical load data to infer the future electricity load. Generally, it is famously known as the Box–Jenkins autoregressive integrated moving average (ARIMA) [7–13], which is theoretically based on univariate time sequences. Christianse [14] and Park et al. [11] design exponential smoothing models by Fourier series transformation for electricity load forecasting. Hence, many researchers consider related factors, such as seasonal temperature and day type, in load forecasting models. Mbamalu and El-Hawary [15] propose multiplicative autoregressive (AR) models that considered seasonal factors in load forecasting. The analytical results show that the forecasting accuracy of the proposed models outperformed the univariate AR model. Douglas et al. [16] consider verifying the impacts of temperature on the forecasting model. The authors combine Bayesian estimation with a dynamic linear model for load forecasting. The experimental results demonstrated that the presented model is suitable for forecasting load under imperfect weather information. Sadownik and Barbosa [17] propose dynamic nonlinear models for load forecasting. The main disadvantage of these methods is that they become time-consuming to compute as the number of variables increases. Azadeh et al. [18] employ fuzzy system to provide an ideal rule base to determine which type of ARMA models should be used; the results also indicate that the integrated approach outperforms those novel intelligent computing models. Wang et al. [19] propose hybrid ARMAX (autoregressive and moving average with exogenous variables) model with particle swarm optimization to efficiently solve the problem of trapping into local minimum which is caused by exogenous variable (e.g., weather condition). Their results also reveal that the proposed approach has superior forecasting accuracy.

To achieve the accuracy of load forecasting, state space and Kalman filtering technologies, developed to reduce the difference between actual loads and prediction loads (random error), are employed in load forecasting model. This approach introduces the periodic component of load as a random process. It requires historical data more than 3–10 years to construct the periodic load variation and to estimate the dependent variables (load or temperature) of power system [20–22]. Moghram and Rahman [23] proposed a model based on this technique and verified that the proposed model outperforms another four forecasting methods (multiple linear regression, time series, exponential smoothing, and knowledge-based approach). Similarly, Park et al. [11] proposed a load forecasting model based on the state space and Kalman filtering technology and also showed that their model out-performed other methods. The disadvantage of these methods is that they are difficult to avoid the observation noise in the forecasting process, especially multivariable

considered. Recently, Al-Hamadi and Soliman [24] employ fuzzy rule-based logic, by utilizing a moving window of current values of weather data as well as the recent past history of load and weather data, to recursively estimate the optimal fuzzy parameters for each hour load of the day. Amjady [25] proposes hybrid model of the forecast-aided state estimator (FASE) and the multilayer perceptron (MLP) neural network to forecast short-term bus load of power systems. The proposed hybrid model has been examined on a real power system, and the results show that the hybrid method has better prediction accuracy than the other models, such as MLP, FASE, and the periodic autoregression (PAR) model.

The regression approach is another popular model for forecasting electricity load. Regression models construct the cause–effect relationships between electricity load and the independent variables. The most popular model is linear regression model, proposed by Asbury [26]; he considers the "weather" variable to explain the electric load. Meanwhile, Papalexopoulos and Hesterberg [27] add the factors of "holiday" and "temperature" into their proposed model. The proposed model uses the weighted least square method to obtain robust parameter estimation encountering with the heteroskedasticity. Furthermore, Soliman et al. [28] propose a multivariate linear regression model for load forecasting, which includes temperature and wind cooling/humidity factors. The empirical results indicate that the proposed model outperforms the harmonic model as well as the hybrid model. Similarly, Mirasgedis et al. [29] also incorporate weather meteorological variables, such as relative humidity, heating, and cooling degree days to forecast electricity demand in Greece. In contrast, Mohamed and Bodger [30] employ economic and geographic variables (such as GDP, electricity price, and population) to forecast electricity consumption in New Zealand. In these models, the dependent variables are generally decomposed into weather insensitive and weather sensitive components [4, 11, 31]. However, these models are all based on linear assumption, that is, these independent variables cannot be well justified due to nonlinear relationships among variables. Therefore, in the recent years, Tsekouras et al. [32] introduce a nonlinear multivariable regression approach to forecast annual load, which considers correlation analysis with weighting factors to select appropriate input variables. Asber et al. [33] employ kernel regression model to establish a relationship among past, current, and future temperatures and the system loads to forecast the load in the Hydro-Quebec distribution network. A set of past load history comprising of weather information and load consumption is used. The paper proposes a class of flexible conditional probability models and techniques for classification and regression problems. A group of regression models is used, each one focusing on consumer classes characterizing specific load behavior. Numerical investigations show that the suggested technique is an efficient way of computing forecast statistics.

1.2 Artificial Intelligent Technology for Electric Load Forecasting

Recently, lots of researches have attempted to apply artificial intelligence techniques to improve the accuracy of electric load forecasting models. Knowledge-based expert system (KBES) and artificial neural networks (ANNs) are the popular representatives. Rahman and Bhatnagar [34] present a KBES model for electricity load forecasting. They construct new rules based on received information, including daily temperature, day type, and load from the previous day. The characteristic feature of this approach is rule based, which implied that the system transformed new rule from received information. In other words, this approach is derived from training rules and transformed the information into mathematical equations; the so-called expert capability is training by the existence presuming and will significantly increase the forecasting accuracy [34–36]. Recently, applying fuzzy inference system and fuzzy theory in load forecasting has also received attentions; Ying and Pan [37] introduce adaptive network fuzzy inference system (ANFIS), by looking for the mapping relation between the input and output data to determine the optimal distribution of membership functions, to forecast regional load. Pai [38] and Pandian et al. [39] all employ fuzzy approaches to get superior performance in terms of load forecasting.

Meanwhile, many researches also have tried to apply ANNs to improve the load forecasting accuracy level. Dillon et al. [40] use adaptive pattern recognition and self-organizing techniques for short-term load forecasting. Dillon et al. [41] present a three-layered feedforward adaptive neural network to forecast short-term load. Their proposed model is trained by back-propagation neural network. The proposed model is applied to real data from a power system and is distinguished providing superior comparative results with other methods are given. In the meanwhile, Park et al. [42] propose a 3-layer back-propagation neural network to daily load forecasting problems. The inputs include three indices of temperature: average, peak, and lowest loads. The outputs are peak loads. The proposed model outperforms the regression model and the time series model in terms of forecasting accuracy index and mean absolute percent error (MAPE). Moreover, Ho et al. [43] develop an adaptive learning algorithm for forecasting the electricity load in Taiwan. The numerical results demonstrate that the proposed algorithm converges faster than the traditional back-propagation learning method. Novak [44] applies radial basis function (RBF) neural networks to forecast electricity load. The analytical results indicate that the RBF network is at least 11 times faster and more reliable than the back-propagation neural networks. Darbellay and Slama [45] apply ANNs to predict the Czech electricity load. The experimental results show that the proposed ANN model outperforms the ARIMA model in terms of normalized mean square error. Abdel-Aal [46] proposes an abductive network to conduct 1 h ahead load forecasts for a 5-year period. The proposed model achieves extremely promising results based on the measurement of mean absolute percent error. Hsu and Chen [47] employ back-propagation neural networks to forecast the regional load in Taiwan. The experiment results show that the artificial neural network approach outperforms the regression models. Recently, Kandil et al. [48] apply

ANNs for short-term load forecasting using real load and weather data from the Hydro-Quebec databases where three types of variables are used as inputs to the neural network. Their proposed model demonstrates ANNs capabilities in load forecasting without the use of load history as an input. In addition, only temperature (from weather variables) is used, in this application, where results show that other variables like sky condition (cloud cover) and wind velocity have no serious effect and may not be considered in the load forecasting procedure. Applications of hybrid ANNs model with statistical methods or other intelligent approaches have received a lot of attention, such as hybrid with Bayesian inference [49, 50], self-organizing map [51, 52], wavelet transform [53, 54], particle swarm optimization [55], and dynamic mechanism [56].

1.3 Support Vector Regression for Electric Load Forecasting

Proposed by Vapnik [57], support vector machines (SVMs) are one of the significant developments in overcoming shortcomings of ANNs mentioned above. Rather than most of the traditional neural network models by implementing the empirical risk minimization (ERM) principle to minimize the training error, SVMs apply the structural risk minimization (SRM) principle to minimize an upper bound on the generalization error. SVMs can theoretically guarantee to achieve the global optimum, instead of trapping local optimum like ANN models. Thus, the solution of a nonlinear problem in the original lower dimensional input space could be equivalent to solving a linear-constrained quadratic programming problem and find its linear solution in the higher dimensional feature space. Originally, SVMs have found wide application in the field of pattern recognition, bioinformatics, and other artificial intelligence relevant applications. In addition, with introduction of Vapnik's ε-insensitive loss function, SVMs have been extended to solve nonlinear regression estimation problems, which are so-called support vector regression (SVR). SVR has been successfully employed to solve forecasting problems in many fields, such as financial time series forecasting [58–65], production value forecasting of machinery industry [66, 67], software reliability forecasting [68], atmospheric science forecasting [69–72], tourism forecasting [73, 74], and so on. Meanwhile, SVR model has also been successfully applied to forecast electric load [75–81]. Cao [58] uses the SVMs experts for time series forecasting. The generalized SVMs experts contain a two-stage neural network architecture. The numerical results indicate that the SVMs experts are capable of outperforming the single SVM models in terms of generalization comparison. Cao and Gu [59] propose a dynamic SVMs model to deal with nonstationary time series problems. Experimental results show that the dynamic SVM model outperforms standard SVMs in forecasting nonstationary time series. Meanwhile, Tay and Cao [60] present C-ascending SVMs to model nonstationary financial time series. Experimental results show that the C-ascending SVMs with actually ordered sample data consistently perform better than standard SVMs. Tay and Cao [61] use SVMs in forecasting financial time series. The numerical results indicate that the SVMs are superior to the multilayer back-propagation neural network in financial time series forecasting. Hong and Pai [68] apply SVR in forecasting rainfall during the period while typhoon attacks Taiwan. The experimental results indicate that SVR outperforms other alternative forecasting models, such as Holt–Winters (HW) model, seasonal Holt and Winters' linear exponential smoothing (SHW) model, and recurrent neural network (RNN) model. Hong and Pai [67] apply SVMs to predict engine reliability. Their experimental results indicate that SVMs outperform Duane model, ARIMA model, and general regression neural networks model. Hong et al. [73] propose a multifactor support vector regression model to forecast Taiwanese demand for travel to Hong Kong from 1967 to 1996. They indicate that the proposed SVRCGA model outperforms BP model, FF model, Holt's model, MA model, naïve model, and multiple regression model. For electric load forecasting, Chen et al. [75] are the pioneers for proposing a SVM model, which is the winning entry of a competition aiming at midterm load forecasting (predicting daily maximum load of the next 31 days) organized by EUNITE network in 2001, to solve the problem. They discuss in detail how the SVM, a new learning technique, is successfully applied to load forecasting. Pai and Hong [80] employ the concepts of Jordan recurrent neural networks to construct recurrent SVR model in Taiwan regional long-term load forecasting. In addition, they use genetic algorithms to determine approximate optimal parameters in the proposed RSVMG model. They conclude that RSVMG outperforms other models, such as SVMG, ANN, and regression models. Similarly, Pai and Hong [81] also propose a hybrid model of SVR and simulated annealing (SA) algorithms to forecast Taiwan's long-term electric load, in which SA is employed to select approximate optimal parameters in the proposed SVMSA model. Conclusively, they indicate that SVMSA is superior to ARIMA and GRNN models in terms of MAPE, MAD, and NRMSE.

The empirical results indicated that selection of the three parameters, C (to trade off the training errors and large weights), ε (the width of the insensitive loss function), and σ (the parameter of Gaussian kernel function), in an SVR model influences the forecasting accuracy significantly. Although numerous publications in the literature have given some recommendations on appropriate setting of SVR parameters [82], however, those approaches do not simultaneously consider the interaction effects among the three parameters. There is no general consensus and many contradictory opinions. It is feasible to employ optimization solving procedure to obtain suitable parameters combination, such as minimizing the objective function describing the structural risk mentioned above. Evolutionary algorithms, such as genetic algorithm, simulated annealing algorithms, immune algorithms, particle swarm optimization, and tabu search, are the very candidates to be employed to determine appropriate parameter values. However, evolutionary algorithms almost lack knowledge memory or storage functions which would be either time-consuming or inefficient in searching the suitable parameters (i.e., being premature convergent or being trapped in local optimum). Therefore, it is necessary to consider some feasible approaches, such as hybridizing or combining other potential technologies, to overcome the premature convergent problems.

1.4 Feasible Approaches to Improve the Forecasting Accuracy Performance

As mentioned, evolutionary algorithms almost have their theoretical drawbacks, such as lack of knowledge memory or storage functions, time-consuming in training, trapped in local optimum, and so on. Therefore, hybridizing some novel search technique to adjust their internal parameters (e.g., mutation rate, crossover rate, and annealing temperature) to overcome the embedded shortcomings is the feasible improving approach. There are three feasible considerations.

1.4.1 Hybridization of Complementary Evolutionary Algorithms

Firstly, for example, in genetic algorithm (GA), new individuals are generated by the following operators: selection, crossover, and mutation. For all types of objective functions, the generation begins with a binary coding for the parameter set. Based on this special binary coding process, GA is able to solve some specified problems which are not easily solved by traditional algorithms. GA can empirically provide a few best-fitted offsprings from the whole population; however, after some generations, due to low diversity of the population, it might lead to a premature convergence. Similarly, simulated annealing (SA) is a generic probabilistic search technique that simulates the material physical process of heating and controlled cooling. Each step of SA attempts to replace the current state by a random move. The new state may then be accepted with a probability that depends both on the difference between the corresponding function values and also on a global parameter, temperature. Thus, SA has some institution to reach more ideal solutions. However, SA costs lots of computation time in annealing process. To improve premature convergence and to receive more suitable objective function values, it is necessary to find some effective approach to overcome these drawbacks from GA to SA. Hybridization of genetic algorithm with simulated annealing (GA–SA) algorithm is an innovative trial by applying the superior capability of SA algorithm to reach more ideal solutions and by employing the mutation process of GA to enhance searching process. GA–SA algorithm has been applied to the fields of system design [83], system and network optimization [84], continuous-time production planning [85], and electrical power districting problem [86]. Furthermore, due to easy implementation process and special mechanism to escape from local optimum [87], chaos and chaos-based searching algorithms have received intense attentions [88, 89]. Applications of chaotic sequence to carefully expand variable searching space, that is, let variable travel ergodically over the searching space, are more and more popular to be employed in evolutionary computation fields.

1.4.2 *Hybridization of Chaos/Cloud Theories with Evolutionary Algorithms*

Secondly, several disadvantages embedded in these evolutionary algorithms are required to be improved to get more satisfied performance. For example, based on the operation procedure of SA, subtle and skillful adjustment in the annealing schedule is required, such as the size of the temperature steps during annealing. Particularly, the temperature of each state is discrete and unchangeable, which does not meet the requirement of continuous decrease in temperature in actual physical annealing processes. In addition, SA is easy to accept deteriorate solution with high temperature, and it is hard to escape from local minimum trap with low temperature [90]. To overcome these drawbacks of SA, the cloud theory is considered. Cloud theory is a model of the uncertainty transformation between quantitative representation and qualitative concept using language value [91]. It is successfully used in intelligence control [92, 93], data mining [94], spatial analysis [95], intelligent algorithm improvement [96], and so on. Based on the operation procedure of SA, subtle and skillful adjustments in the annealing schedule are required, such as the size of the temperature steps during annealing, the temperature range, and the number of restarts and redirection of the search. The annealing process is like a fuzzy system in which the molecules move from large scale to small scale randomly as the temperature decreases. In addition, due to its Monte Carlo scheme and lack of knowledge memory functions, time-consuming is also an another boring problem. Author has tried to employ chaotic simulated annealing (CSA) algorithm, to overcome these shortcomings, in which the transiently chaotic dynamics are temporarily generated for foraging and self-organizing, then gradually vanished with autonomous decreasing of the temperature, and are accompanied by successive bifurcations and converged to a stable equilibrium. Therefore, CSA has significantly improved the randomization of Monte Carlo scheme, has controlled the convergent process by bifurcation structures instead of stochastic “thermal” fluctuations, and eventually performed efficient searching including a global optimum state. However, as mentioned that the temperature of each state is discrete and unchangeable, which does not meet the requirement of temperature continuously decrease in actual physical annealing processes. Even some temperature annealing function is exponential in general, the temperature is gradually fallen with a fixed value in every annealing step, and the changing process of temperature between two neighbor steps is not continuous. This phenomenon also appears while other types of temperature update functions are implemented, such as arithmetical, geometrical, or logarithmic one. In the cloud theory, by introducing the Y condition normal cloud generator to the temperature generation process, it can randomly generate a group of new values that distribute around the given value like “cloud.” Let the fixed temperature point of each step become a changeable temperature zone, the temperature of each state generation in every annealing step is chosen randomly, the course of temperature changing in the whole annealing process is nearly continuous and fits the physical annealing process better. Therefore, based on chaotic sequence and cloud theory, the CCSA is employed to replace the stochastic “thermal” fluctuations control from traditional SA, to enhance the continuously physical temperature annealing process from CSA. The cloud theory can realize the transformation between a qualitative concept in words and its numerical representation. It is able to be employed to avoid problems mentioned above.

1.4.3 Combination of Recurrent/Seasonal Mechanisms with Evolutionary Algorithms

Thirdly, the concepts of combined or hybrid models also deserve to be considered. Please notice that the so-called hybrid model means that some process of the former model is integrated into the process of the latter one, for example, hybridizing A and B implies that some processes of A are controlled by A and some are by B. On the other hand, for the so-called combined model, it only indicated that the output of the former model is then the input of the latter one; therefore, the classification results from combined models will be superior to single model. The combined models are employed to further capture more data pattern information from the analyzed data series. For example, inspired by the concept of recurrent neural networks (RNNs) that every unit is considered as an output of the network and the provision of adjusted information as input in a training process [97], the recurrent learning mechanism framework is also combined into the original analyzed model. For a feedforward neural network, links may be established within layers of a neural network. These types of networks are called recurrent neural networks. RNNs are extensively applied in time series forecasting. Jordan [98] proposes a recurrent neural network model (Fig. 1.1) for controlling robots. Elman [99] develops a recurrent neural network model (Fig. 1.2) to solve linguistics problems. Williams and Zipser [100] present a recurrent network model (Fig. 1.3) to solve nonlinear adaptive filtering and pattern recognition problems. These three models mentioned all consist of multilayer perceptron (MLP) with a hidden layer. Jordan networks have a feedback loop from the output layer with past values to an additional input, namely, "context layer." Then, output values from the context layer are fed back into the hidden layer. Elman networks have a feedback loop from the hidden layer to the context layer. In Williams and Zipser networks, nodes in the hidden layer are fully connected to each other. Both Jordan and Elman networks include an additional information source from the output layer or the hidden layer. Hence, these models use mainly past information to capture detailed information. Williams and Zipser networks take much more information from the hidden layer and back into themselves. Therefore, Williams and Zipser networks are sensitive when models are implemented [101]. For another combined model, on the other hand, some data series sometimes reveals a seasonal tendency due to cyclic economic activities or seasonal nature hour to hour, day to day, week to week, month to month, and season to season, such as hourly peak in a working day, weekly peak in a business week, and monthly peak in a demand planned year. In order to excellently deal with cyclic/seasonal trend data series, some useful trial, for example, seasonal mechanism [102, 103], also received some intentions.

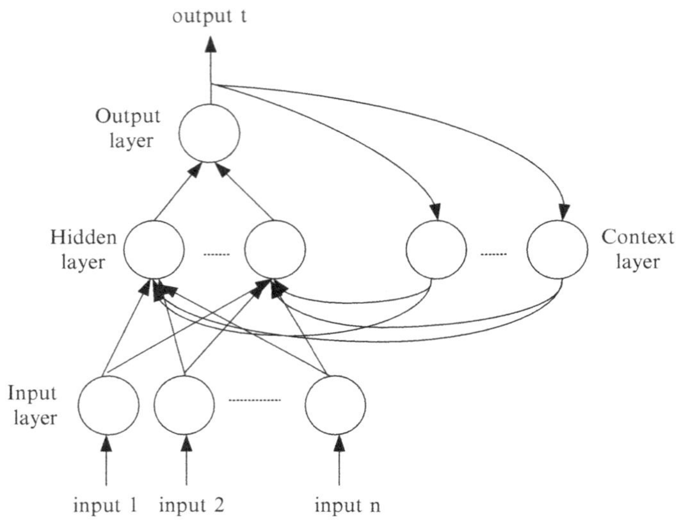

Fig. 1.1 Network diagram created from Jordan's definition

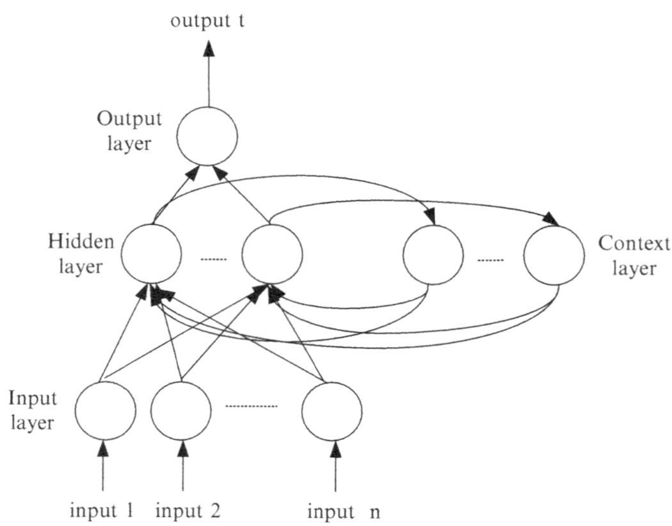

Fig. 1.2 Network diagram created from Elman's definition

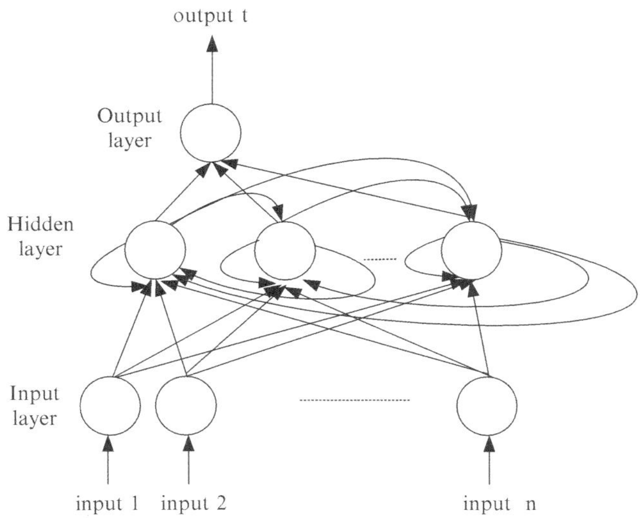

Fig. 1.3 Network diagram created from Williams and Zipser's definition

1.4.4 Summary: Electric Load Forecasting Support System (ELFSS)

Based on the discussions above, it will also become the research mainstream in SVR-based electric load forecasting, which is also the principal purpose of this book, to guide researchers how to employ alternative ways (proper evolutionary algorithms) in parameter determination while new electric load forecasting model is constructing, that is, the importance and necessity of the electric load forecasting support systems. This is because that for any forecasting model, the most important problem is how to catch the data pattern and apply the learned patterns or rules to forecast, that is, the key successful factor is how to suitably look for data pattern. The data patterns could be classified into three categories: (1) fluctuation, changing violently according to policy or herding behaviors of investors; (2) regular pattern, trends (electric load increasing or decreasing annually) or seasonality/cyclic (peak electric load in summer and winter); and (3) noise, accidental events (e.g., 9/11 event, SARS event) or man-made events (e.g., product promotion event). However, each model itself has excelled ability to catch specific data pattern. For example, exponential smoothing and ARIMA models focus on strict increasing (or decreasing) time series data, that is, linear pattern, even though they have seasonal modification mechanism to analyze seasonal (cyclic) change; due to artificial learning function being able to adjust the suitable training rules, ANN model is excelled only if historical data pattern has been learned, it is lacks of systematic explanation how the accurate forecasting results are obtained; SVR model could acquire superior performance only if proper parameter determination search algorithms.

Therefore, it is essential to construct an inference system to collect the characteristic rules to determine the data pattern category. Then, it should assign appropriate approach to implement forecasting: for (1) ARIMA or exponential smoothing approaches, the only work is to adjust their differential or seasonal parameters, and (2) for ANN or SVR models, the forthcoming problem is how to determine best parameter combination (numbers of hidden layer, units of each layer, learning rate or Gaussian σ, model flatness C, ε-insensitive) to acquire superior forecasting performance. Particularly, for the focus of this book, in order to determine the most proper parameter combination (σ, C, and ε), a series of evolutionary algorithms should be employed to test which data pattern is familiar with, such as genetic algorithms (GA), simulated annealing algorithms (SA), ant colony optimization (ACO), tabu search (TA), immune algorithm (IA), and particle swarm optimization algorithm (PSO). Based on experimental findings, those evolutionary algorithms themselves also have merits and drawbacks, for example, GA and IA could handle excellently in regular trend data pattern (real number) [80], SA excelled in fluctuation or noise data pattern (real number) [68, 81], TA is good in regular cyclic data pattern (real number) [104], and ACO is well done in integer number searching.

As aforementioned, it is possible to propose an intelligent forecasting support system to improve the usage efficiency of evolutionary algorithms, chaos/cloud theories, and recurrent/seasonal mechanisms hybridized in an SVR load forecasting model, namely, electric load forecasting support system (ELFSS). The main flowchart of the ELFSS suggested in this conclusion is given in Fig. 1.4. Firstly, employ fuzzy logic to construct the inference system to preprocess the time series data and find out or define the characteristic rules set of data pattern, such as linear, logarithmic, inverse, quadratic, cubic, compound, power, growth, and exponential. Secondly, filter the original electric load data by those data pattern rules set and then recognize the appropriate data pattern (fluctuation, regular, or noise). The recognition decision rules should include two principles (1) the change rate of two continuous electric load data and (2) the decreasing or increasing trend of the change rate, that is, behavior of the approached curve. Finally, decide appropriate evolutionary algorithm (including hybrid evolutionary algorithms) to be hybridized into an SVR model; in addition, to avoid trapping in local optimum, suitable chaos or cloud theory and appropriate (recurrent or seasonal) mechanism could be further hybridized or combined with associated evolutionary algorithms into these SVR-based forecasting models (such as CGA, CSA, CTA, CIA, CACO, and CPSO in Fig. 1.4).

1.5 Structure of This Book

In this book, different techniques used in the past decades are employed to construct the electric load forecasting models, including ARIMA, SARIMA, HW, SHW, GRNN, and BPNN models; chaos/cloud theories; and recurrent/seasonal mechanisms. The book contains six chapters:

Chapter 1, "Introduction." This chapter introduces the background of electric load forecasting, traditional approaches, artificial intelligent technology, SVR for electric

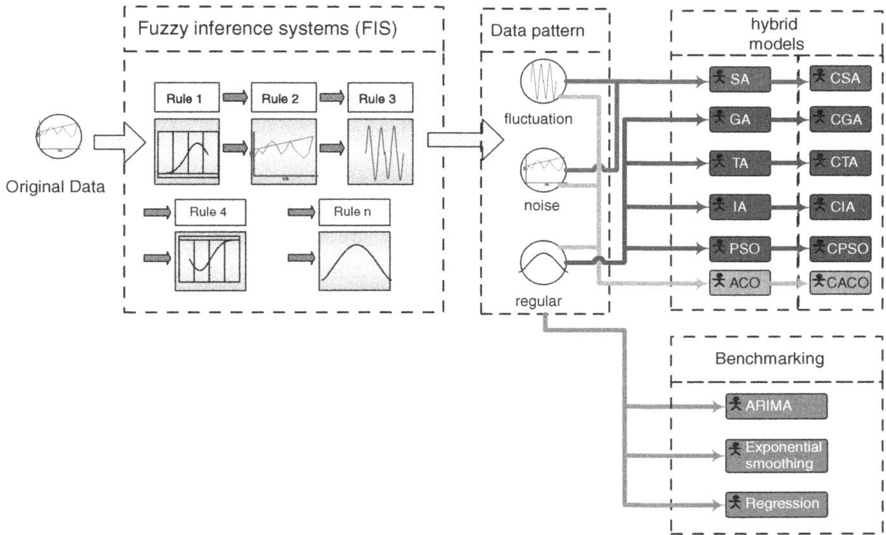

Fig. 1.4 The electric load forecasting support system (ELFSS)

load forecasting, and some feasible improvements of forecasting accuracy, to help the reader understand the very issue of electric load forecasting in this book and the current development tendency and shortcomings; in addition, some critical arrangements to improve the forecasting accuracy level are also discussed. In this chapter, readers will study the basic concepts of the electric load forecasting and associate forecasting technologies, including ARIMA, exponential smoothing, state space/Kalman filtering, regression, knowledge-based expert system (KBES), artificial neural networks (ANNs), fuzzy theory, support vector regression, and so on. The reader will also review these technologies proposed during the past decades for electric load forecasting. A brief discussion for each model is given in this chapter. Superiorities and shortcomings of each model are also taken into account and discussed.

Chapter 2, "Modeling for Energy Demand Forecasting." This chapter introduces different basic energy demand forecasting models, which will be employed in Chaps. 3–5. Electric load forecasting methods can be classified in three categories (1) traditional approaches, including Box–Jenkins autoregressive integrated moving average (ARIMA) model, autoregressive and moving average with exogenous variables (ARMAX) model, seasonal ARIMA (SARIMA) model, exponential smoothing models (including Holt–Winters model (HW) and seasonal Holt and Winters' linear exponential smoothing (SHW)), state space/Kalman filtering model, and linear regression model; (2) artificial intelligent approaches, including knowledge-based expert system (KBES) model, artificial neural networks (ANNs) model, and fuzzy inference system model; and (3) support vector regression (SVR) model and its related hybrid/combined models. These models are classified according to the basis of the forecasting technological development tendency, evolved from mathematical relationship model (e.g., statistics-based model) to application of artificial intelligent model (e.g., ANNs model) and eventually to hybridization of statistical model and artificial intelligent model (e.g., SVR model). Of course, the classifications are not unique, and the classification based on the technological evolution is not always suitable for another. Based on this classification, interested readers can be inspired to propose another new model to receive more accurate electric load forecasting performance. Each model has its outstanding advantages compared with other models due to its theoretical innovation and also has its embedded theoretical limitations; thus, it always has the potential to be improved by hybridizing or combining with other novel approaches. Seven representative models are introduced, namely, ARIMA, SARIMA, Holt–Winters (HW), seasonal HW (SHW), general regression neural network (GRNN), back-propagation neural networks (BPNN), and SVR models.

Chapter 3, "Evolutionary Algorithms in SVR's Parameters Determination." As mentioned, the traditional determination of three parameters does not guarantee to improve forecasting accuracy level, because it is unable to set up more suitable initial values of parameters in the initial step and unable to simultaneously consider the interaction effects among three parameters to efficiently find out the near-optimal solution for large-scale data set. Therefore, evolutionary algorithms are employed to conduct intelligent searching around the solution range to determine suitable parameter combination by minimizing the objective function describing the structural risk of an SVR model. This chapter introduces several representative evolutionary algorithms, such as genetic algorithm (GA), simulated annealing (SA) algorithm, hybrid GA with SA (GA–SA) algorithm, particle swarm optimization (PSO) algorithm, ant colony optimization (ACO) algorithm, artificial bee colony (ABC) algorithm, and immune algorithm (IA), used in an SVR forecasting model to determine suitable parameter combination to receive improved forecasting accuracy level.

Chapter 4, "Chaos/Cloud Theories to Avoid Trapping into Local Optimum." As demonstrated in Chap. 3, these different evolutionary algorithms, including genetic algorithm (GA), simulated annealing (SA) algorithm, hybrid GASA algorithm, particle swarm optimization (PSO) algorithm, continuous ant colony optimization (CACO) algorithm, artificial bee colony (ABC) algorithm, and immune algorithm (IA), are employed to determine suitable parameter combination of an SVR-based electric load forecasting model. These forecasting results indicate that almost all SVR-based models with different evolutionary algorithms are superior to other competitive forecasting models (including ARIMA, HW, GRNN, and BPNN models); however, these algorithms almost lack knowledge memory or storage mechanisms which would be either time-consuming or inefficient in searching the suitable parameters, that is, premature convergence (being trapped in local optimum). This chapter introduces that hybrid chaos theory with evolutionary algorithms can overcome the shortcomings of trapping local optimum to improve forecasting performance.

Chapter 5, "Recurrent/Seasonal Mechanism to Improve the Accurate Level of Forecasting." As demonstrated in Chap. 4, these different hybrid chaotic evolutionary algorithms, including chaotic genetic algorithm (CGA), chaotic simulated annealing (CSA) algorithm, chaotic cloud simulated annealing (CCSA) algorithm, chaotic GASA (CGASA) algorithm, chaotic particle swarm optimization (CPSO) algorithm, chaotic ant swarm (CAS) algorithm, chaotic artificial bee colony (CABC) algorithm, and chaotic immune algorithm (CIA), employed to determine suitable parameter combination of an SVR-based electric load forecasting model. These forecasting results indicate that almost all SVR-based models with different hybrid chaotic-evolutionary algorithms are superior to other competitive forecasting models (including ARIMA, GRNN, and TF-ε-SVR-SA models). However, these hybrid chaotic-evolutionary algorithms do not provide satisfactory forecasting performance (well fitting the actual fluctuation tendency), even though their forecasting accuracy receives significant level. To improve the fitting effects for each SVR-chaotic/cloud-evolutionary algorithm-based model, this chapter introduces two combined mechanisms (recurrent mechanism or seasonal mechanism) to significantly improve the fitting effects with the actual fluctuation tendency.

References

1. Gross G, Galiana FD (1987) Short term load forecasting. Proc IEEE 75:1558–1573. doi:10.1109/PROC.1987.13927
2. Ranaweera DK, Karady GG, Farmer RG (1997) Economic impact analysis of load forecasting. IEEE Trans Power Syst 12:1388–1392. doi:10.1109/59.630486
3. Douglas AP, Breipohl AM, Lee FN, Adapa R (1998) Risk due to load forecast uncertainty in short term power system planning. IEEE Trans Power Syst 13:1493–1499. doi:10.1109/59.736296
4. Bunn DW, Farmer ED (1985) Comparative models for electrical load forecasting. Wiley, New York
5. Bunn DW (2000) Forecasting loads and prices in competitive power markets. Proc IEEE 88:163–169. doi:10.1109/5.823996
6. Amjady N, Keynia F (2009) Short-term load forecasting of power systems by combination of wavelet transform and neuro-evolutionary algorithm. Energy 34:46–57
7. Box GEP, Jenkins GM (1970) Time series analysis, forecasting and control. Holden-Day, San Francisco
8. Saab S, Badr E, Nasr G (2001) Univariate modeling and forecasting of energy consumption: the case of electricity in Lebanon. Energy 26:1–14. doi:10.1016/S0360-5442(00)00049-9
9. Chen JF, Wang WM, Huang CM (1995) Analysis of an adaptive time-series autoregressive moving-average (ARMA) model for short-term load forecasting. Electric Power Syst Res 34:187–196. doi:10.1016/0378-7796(95)00977-1
10. Wang H, Schulz NN (2006) Using AMR data for load estimation for distribution system analysis. Electric Power Syst Res 76:336–342. doi:10.1016/j.epsr.2005. 08.003
11. Park JH, Park YM, Lee KY (1991) Composite modeling for adaptive short-term load forecasting. IEEE Trans Power Syst 6:450–457. doi:10.1109/59.76686
12. Abdel-Aal RE, Al-Garni AZ (1997) Forecasting monthly electric energy consumption in eastern Saudi Arabia using univariate time-series analysis. Energy 22:1059–1069. doi:10.1016/S0360-5442(97)00032-7
13. Chavez SG, Bernat JX, Coalla HL (1999) Forecasting of energy production and consumption in Asturias (northern Spain). Energy 24:183–198. doi:10.1016/S0360-5442(98)00099-1
14. Christianse WR (1971) Short term load forecasting using general exponential smoothing. IEEE Trans Power Apparatus Syst PAS-90:900–911. doi:10.1109/TPAS.1971.293123
15. Mbamalu GAN, El-Hawary ME (1993) Load forecasting via suboptimal seasonal autoregressive models and iteratively reweighted least squares estimation. IEEE Trans Power Syst 8:343–348. doi:10.1109/59.221222

16. Douglas AP, Breipohl AM, Lee FN, Adapa R (1998) The impact of temperature forecast uncertainty on Bayesian load forecasting. IEEE Trans Power Syst 13:1507–1513. doi:10.1109/59.736298
17. Sadownik R, Barbosa EP (1999) Short-term forecasting of industrial electricity consumption in Brazil. J Forecast 18:215–224. doi:10.1002/(SICI)1099-131X (199905)18:3<215::AID-FOR719>3.0.CO;2-B
18. Azadeh A, Saberi M, Ghaderi SF, Gitiforouz A, Ebrahimipour V (2008) Improved estimation of electricity demand function by integration of fuzzy system and data mining approach. Energy Convers Manage 49:2165–2177. doi:10.1016/j.enconman.2008.02.021
19. Wang B, Tai NL, Zhai HQ, Ye J, Zhu JD, Qi LB (2008) A new ARMAX model based on evolutionary algorithm and particle swarm optimization for short term load forecasting. Electric Power Syst Res 78:1679–1685. doi:10.1016/j.epsr. 2008.02.009
20. Brown RG (1983) Introduction to random signal analysis and Kalman filtering. Wiley, New York
21. Gelb A (1974) Applied optimal estimation. The MIT Press, Cambridge, MA
22. Trudnowski DJ, McReynolds WL, Johnson JM (2001) Real-time very short-term load prediction for power-system automatic generation control. IEEE Trans Control Syst Technol 9:254–260. doi:10.1109/87.911377
23. Moghram I, Rahman S (1989) Analysis and evaluation of five short-term load forecasting techniques. IEEE Trans Power Syst 4:1484–1491. doi:10.1109/59. 41700
24. Al-Hamadi HM, Soliman SA (2006) Fuzzy short-term electric load forecasting using Kalman filter. IEE Proc Gener Transm Distrib 153:217–227. doi:10.1049/ip-gtd:20050088
25. Amjady N (2007) Short-term bus load forecasting of power systems by a new hybrid method. IEEE Trans Power Syst 22:333–341. doi:10.1109/TPWRS.2006. 889130
26. Asbury C (1975) Weather load model for electric demand energy forecasting. IEEE Trans Power Apparatus Syst PAS-94:1111–1116. doi:10.1109/T-PAS.1975. 31945
27. Papalexopoulos AD, Hesterberg TC (1990) A regression-based approach to short-term system load forecasting. IEEE Trans Power Syst 5:1535–1547. doi:10. 1109/59.99410
28. Soliman SA, Persaud S, El-Nagar K, El-Hawary ME (1997) Application of least absolute value parameter estimation based on linear programming to short-term load forecasting. Int J Electr Power Energ Syst 19:209–216. doi:10.1016/S0142-0615(96)00048-8
29. Mirasgedis S, Safaridis Y, Georgopoulou E, Lalas DP, Moschovits M, Karagiannis F, Papakonstantinou D (2006) Models for mid-term electricity demand forecasting incorporating weather influences. Energy 31:208–227. doi:10.1016/j.energy.2005.02.016
30. Mohamed Z, Bodger P (2005) Forecasting electricity consumption in New Zealand using economic and demographic variables. Energy 30:1833–1843. doi:10.1016/j.energy.2004.08.012
31. Hyde O, Hodnett PF (1997) An adaptable automated procedure for short-term electricity load forecasting. IEEE Trans Power Syst 12:84–93. doi:10.1109/59. 574927
32. Tsekouras GJ, Dialynas EN, Hatziargyriou ND, Kavatza S (2007) A non-linear multivariable regression model for midterm energy forecasting of power systems. Electric Power Syst Res 77:1560–1568. doi:10.1016/j.epsr.2006.11.003
33. Asber D, Lefebvre S, Asber J, Saad M, Desbiens C (2007) Non-parametric short-term load forecasting. Int J Electr Power Energ Syst 29:630–635. doi:10.1016/j.ijepes.2006.09.007
34. Rahman S, Bhatnagar R (1988) An expert system based algorithm for short-term load forecasting. IEEE Trans Power Syst 3:292–299. doi:10.1109/59.192889
35. Rahman S, Hazim O (1993) A generalized knowledge-based short-term load-forecasting technique. IEEE Trans Power Syst 8:508–514. doi:10.1109/59. 260833
36. Chiu CC, Kao LJ (1997) Combining a neural network with a rule-based expert system approach for short-term power load forecasting in Taiwan. Expert Syst Appl 13:299–305. doi:10.1016/S0957-4174(97)00048-1

37. Ying LC, Pan MC (2008) Using adaptive network based fuzzy inference system to forecast regional electricity loads. Energy Convers Manage 49:205–211. doi:10.1016/j.enconman.2007.06.015
38. Pai PF (2006) Hybrid ellipsoidal fuzzy systems in forecasting regional electricity loads. Energy Convers Manage 47:2283–2289. doi:10.1016/j.enconman.2005. 11.017
39. Pandian SC, Duraiswamy K, Rajan CCA, Kanagaraj N (2006) Fuzzy approach for short term load forecasting. Electric Power Syst Res 76:541–548. doi:10.1016/j.epsr.2005.09.018
40. Dillon TS, Morsztyn K, Phua, K (1975) Short term load forecasting using adaptive pattern recognition and self organizing techniques. In: Proceedings of the fifth world power system computation conference (PSCC-5), Cambridge, UK
41. Dillon TS, Sestito S, Leung S (1991) Short term load forecasting using an adaptive neural network. Int J Electr Power Energ Syst 13:186–192. doi:10.1016/0142-0615(91)90021-M
42. Park DC, El-Sharkawi MA, Marks RJ II, Atlas LE, Damborg MJ (1991) Electric load forecasting using an artificial neural network. IEEE Trans Power Syst 6:442–449. doi:10.1109/59.76685
43. Ho KL, Hsu YY, Yang CC (1992) Short-term load forecasting using a multilayer neural network with an adaptive learning algorithm. IEEE Trans Power Syst 7:141–149. doi:10.1109/59.141697
44. Novak B (1995) Superfast auto-configuring artificial neural networks and their application to power systems. Electric Power Syst Res 35:11–16. doi:10.1016/0378-7796(95)00980-9
45. Darbellay GA, Slama M (2000) Forecasting the short-term demand for electricity—do neural networks stand a better chance. Int J Forecast 16:71–83. doi:10.1016/S0169-2070(99)00045-X
46. Abdel-Aal RE (2004) Short-term hourly load forecasting using abductive networks. IEEE Trans Power Syst 19:164–173. doi:10.1109/TPWRS.2003. 820695
47. Hsu CC, Chen CY (2003) Regional load forecasting in Taiwan: applications of artificial neural networks. Energy Convers Manage 44:1941–1949. doi:10.1016/S0196-8904(02)00225-X
48. Kandil N, Wamkeue R, Saad M, Georges S (2006) An efficient approach for short term load forecasting using artificial neural networks. Int J Electr Power Energ Syst 28:525–530. doi:10.1016/j.ijepes.2006.02.014
49. Saini LM (2008) Peak load forecasting using Bayesian regularization, resilient and adaptive backpropagation learning based artificial neural networks. Electr Power Syst Res 78: 1302–1310. doi:10.1016/j.epsr.2007.11.003
50. Lauret P, Fock E, Randrianarivony RN, Manicom-Ramsamy JF (2008) Bayesian neural network approach to short time load forecasting. Energy Convers Manage 49:1156–1166. doi:10.1016/j.enconman.2007.09.009
51. Amin-Naseri MR, Soroush AR (2008) Combined use of unsupervised and supervised learning for daily peak load forecasting. Energy Convers Manage 49:1302–1308. doi:10.1016/j.enconman.2008.01.016
52. Carpinteiro OAS, Leme RC, de Souza ACZ, Pinheiro CAM, Moreira EM (2007) Long-term load forecasting via a hierarchical neural model with time integrators. Electric Power Syst Res 77:371–378. doi:10.1016/j.epsr.2006.03.014
53. Cao J, Lin X (2008) Study of hourly and daily solar irradiation forecast using diagonal recurrent wavelet neural networks. Energy Convers Manage 49:1396–1406. doi:10.1016/j.enconman.2007.12.030
54. Tai N, Stenzel J, Wu H (2006) Techniques of applying wavelet transform into combined model for short-term load forecasting. Electric Power Syst Res 76:525–533. doi:10.1016/j.epsr.2005.07.003
55. El-Telbany M, El-Karmi F (2008) Short-term forecasting of Jordanian electricity demand using particle swarm optimization. Electric Power Syst Res 78:425–433. doi:10.1016/j.epsr.2007.03.011
56. Ghiassi M, Zimbra DK, Saidane H (2006) Medium term system load forecasting with a dynamic artificial neural network model. Electric Power Syst Res 76:302–316. doi:10.1016/j.epsr.2005.06.010
57. Vapnik V (1995) The nature of statistical learning theory. Springer, New York

58. Cao L (2003) Support vector machines experts for time series forecasting. Neurocomputing 51:321–39. doi:10.1016/S0925-2312(02)00577-5
59. Cao L, Gu Q (2002) Dynamic support vector machines for non-stationary time series forecasting. Intell Data Anal 6:67–83
60. Tay FEH, Cao L (2002) Modified support vector machines in financial time series forecasting. Neurocomputing 48:847–861. doi:10.1016/S0925-2312(01) 00676-2
61. Tay FEH, Cao L (2001) Application of support vector machines in financial time series forecasting. Omega 29:309–317. doi:10.1016/S0305-0483(01)00026-3
62. Huang W, Nakamori Y, Wang SY (2005) Forecasting stock market movement direction with support vector machine. Comput Oper Res 32:2513–2522. doi:10.1016/j.cor.2004.03.016
63. Hung WM, Hong WC (2009) Application of SVR with improved ant colony optimization algorithms in exchange rate forecasting. Contr Cybern 38:863–891
64. Pai PF, Lin CS (2005) A hybrid ARIMA and support vector machines model in stock price forecasting. Omega 33:497–505. doi:10.1016/j.omega.2004.07.024
65. Pai PF, Lin CS, Hong WC, Chen CT (2006) A hybrid support vector machine regression for exchange rate prediction. Int J Inf Manage Sci 17:19–32
66. Pai PF, Lin CS (2005) Using support vector machines to forecast the production values of the machinery industry in Taiwan. Int J Adv Manuf Tech 27:205–210. doi:10.1007/s00170-004-2139-y
67. Hong WC, Pai PF (2006) Predicting engine reliability by support vector machines. Int J Adv Manuf Tech 28:154–161. doi:10.1007/s00170-004-2340-z
68. Pai PF, Hong WC (2006) Software reliability forecasting by support vector machines with simulated annealing algorithms. J Syst Software 79:747–755. doi:10.1016/j.jss.2005.02.025
69. Hong WC, Pai PF (2007) Potential assessment of the support vector regression technique in rainfall forecasting. Water Resour Manage 21:495–513. doi:10.1007/s11269-006-9026-2
70. Mohandes MA, Halawani TO, Rehman S, Hussain AA (2004) Support vector machines for wind speed prediction. Renew Energy 29:939–947. doi:10.1016/j.renene.2003.11.009
71. Pai PF, Hong WC (2007) A recurrent support vector regression model in rainfall forecasting. Hydrol Process 21:819–827. doi:10.1002/hyp. 6323
72. Hong WC (2008) Rainfall forecasting by technological machine learning models. Appl Math Comput 200:41–57. doi:10.1016/j.amc.2007.10.046
73. Hong WC, Dong Y, Chen LY, Wei SY (2011) SVR with hybrid chaotic genetic algorithms for tourism demand forecasting. Appl Soft Comput 11:1881–1890. doi:10.1016/j.asoc.2010.06.003
74. Pai PF, Hong WC (2005) An improved neural network model in forecasting arrivals. Ann Touris Res 32:1138–1141. doi:10.1016/j.annals.2005.01.002
75. Chen BJ, Chang MW, Lin CJ (2004) Load forecasting using support vector machines: a study on EUNITE competition 2001. IEEE Trans Power Syst 19:1821–1830. doi:10.1109/TPWRS.2004.835679
76. Hong WC (2009) Hybrid evolutionary algorithms in a SVR-based electric load forecasting model. Int J Electr Power Energ Syst 31:409–417. doi:10.1016/j.ijepes.2009.03.020
77. Hong WC (2009) Chaotic particle swarm optimization algorithm in a support vector regression electric load forecasting model. Energy Convers Manage 50:105–117. doi:10.1016/j.enconman.2008.08.031
78. Hong WC (2009) Electric load forecasting by support vector model. Appl Math Model 33: 2444–2454. doi:10.1016/j.apm.2008.07.010
79. Hong WC (2010) Application of chaotic ant swarm optimization in electric load forecasting. Energ Policy 38:5830–5839. doi:10.1016/j.enpol.2010.05.033
80. Pai PF, Hong WC (2005) Forecasting regional electric load based on recurrent support vector machines with genetic algorithms. Electric Power Syst Res 74:417–425. doi:10.1016/j.epsr.2005.01.006
81. Pai PF, Hong WC (2005) Support vector machines with simulated annealing algorithms in electricity load forecasting. Energy Convers Manage 46:2669–2688. doi:10.1016/j.enconman.2005.02.004

82. Cherkassky V, Ma Y (2004) Practical selection of SVM parameters and noise estimation for SVM regression. Neural Netw 17:113–126. doi:10.1016/S0893-6080(03)00169-2
83. Shieh HJ, Peralta RC (2005) Optimal in situ bioremediation design by hybrid genetic algorithm-simulated annealing. J Water Resour Plan Manage-ASCE 131:67–78. doi:10.1061/(ASCE) 0733-9496
84. Zhao F, Zeng X (2006) Simulated annealing—genetic algorithm for transit network optimization. J Comput Civil Eng 20:57–68. doi:10.1061/(ASCE)0887- 3801
85. Ganesh K, Punniyamoorthy M (2005) Optimization of continuous-time production planning using hybrid genetic algorithms-simulated annealing. Int J Adv Manuf Tech 26:148–154. doi:10.1007/s00170-003-1976-4
86. Bergey PK, Ragsdale CT, Hoskote M (2003) A simulated annealing genetic algorithm for the electrical power districting problem. Ann Oper Res 121:33–55. doi:10.1023/A:1023347000978
87. Wang L, Zheng DZ, Lin QS (2001) Survey on chaotic optimization methods. Comput Technol Autom 20:1–5
88. Liu B, Wang L, Jin YH, Tang F, Huang DX (2005) Improved particle swarm optimization combined with chaos. Chaos Soliton Fract 25:1261–1271. doi:10. 1016/j.chaos.2004.11.095
89. Cai J, Ma X, Li L, Peng H (2007) Chaotic particle swarm optimization for economic dispatch considering the generator constraints. Energy Convers Manage 48:645–653. doi:10.1016/ j.enconman.2006.05.020
90. Lv P, Lin Y, Zhang J (2009) Cloud theory-based simulated annealing algorithm and application. Eng Appl Artif Intel 22:742–749. doi:10.1016/j.engappai. 2009.03.003
91. Li D, Meng H, Shi X (1995) Membership clouds and membership cloud generators. J Comput Res Dev 32:15–20. doi:CNKI:SUN:JFYZ.0.1995-06-002
92. Li D, Cheung D, Shi X, Ng V (1998) Uncertainty reasoning based on cloud models in controllers. Comput Math Appl 35:99–123. doi:10.1016/S0898-1221 (97)00282-4
93. Zhang F, Fan Y, Shen C, Li D (1999) Intelligent control based membership cloud generators. Acta Aeronautica et Astronautica Sinica 20:89–92
94. Wang S, Li D, Shi W, Li D, Wang X (2003) Cloud model-based spatial data mining. Geogr Inf Sci 9:60–70. doi:10.1080/10824000309480589
95. Wang H, Deng Y (2007) Spatial clustering method based on cloud model. Proceedings of the fourth international conference on fuzzy systems and knowledge discovery, vol 7, pp 272–276. doi:10.1109/FSKD.2007.526
96. Zhu Y, Dai C, Chen W (2005) Adaptive probabilities of crossover and mutation in genetic algorithms based on cloud generators. J Comput Inf Syst 1:671–678
97. Kechriotis G, Zervas E, Manolakos ES (1994) Using recurrent neural networks for adaptive communication channel equalization. IEEE Trans Neural Netw 5:267–278. doi:10.1109/ 72.279190
98. Jordan MI (1986) Attractor dynamics and parallelism in a connectionist sequential machine. In: Proceedings the 8th annual conference of the cognitive science society, Hillsdale, NJ, pp 531–546
99. Elman JL (1990) Finding structure in time. Cogn Sci 14:179–211. doi:10.1207/s15516709cog 1402_1
100. Williams R, Zipser D (1989) A learning algorithm for continually running fully recurrent neural networks. Neural Comput 1:270–280. doi:10.1162/neco. 1989.1.2.270
101. Tsoi AC, Back AD (1994) Locally recurrent globally feed forward networks: a critical review of architectures. IEEE Trans Neural Netw 5:229–239. doi:10. 1109/72.279187
102. Azadeh A, Ghaderi SF (2008) Annual electricity consumption forecasting by neural network in high energy consuming industrial sectors. Energy Convers Manage 49:2272–2278. doi:10.1016/j.enconman.2008.01.035
103. Deo R, Hurvich C, Lu Y (2006) Forecasting realized volatility using a long- memory stochastic volatility model: estimation, prediction and seasonal adjustment. J Econom 131: 29–58. doi:10.1016/j.jeconom.2005.01.003

104. Hong WC, Pai PF, Yang SL, Theng R (2006) Highway traffic forecasting by support vector regression model with tabu search algorithms. In: Proceedings of the IEEE international joint conference on neural networks, Vancouver, Canada, pp 1617–1621. doi:10.1109/IJCNN.2006.246627

Chapter 2
Modeling for Energy Demand Forecasting

As mentioned in Chap. 1, the electric load forecasting methods can be classified in three categories [1–12]:

- Traditional approaches, including Box–Jenkins autoregressive integrated moving average (ARIMA) model, autoregressive and moving average with exogenous variables (ARMAX) model, seasonal autoregressive integrated moving average (SARIMA) model, exponential smoothing models [including Holt–Winters model (HW) and seasonal Holt and Winters' linear exponential smoothing (SHW)], state space/Kalman filtering model, and linear regression model
- Artificial intelligent approaches, including knowledge-based expert system (KBES) model, artificial neural networks (ANNs) model, and fuzzy inference system model
- Support vector regression (SVR) model and its related hybrid/combined models

These models are classified on the basis of the forecasting technological development tendency, evolved from mathematical relationship model (e.g., statistics-based model) to application of artificial intelligent model (e.g., ANNs model) and eventually to hybridizing statistical model and artificial intelligent model (e.g., SVR model). Of course, the classifications are not unique, and the classification based on the technological evolution is not always suitable for another. However, based on this classification, interested readers can be inspired to propose another new model to receive more accurate electric load forecasting performance. Additionally, each model has its outstanding advantages compared with other models due to its theoretical innovation while it has been proposed and also has its embedded theoretical limitations; thus, it always has the potential to be improved by hybridizing or combining with other novel approaches.

This book is focused on SVR model, SVR with hybrid evolutionary algorithms, and SVR with combined mechanisms; therefore, to be based on the same comparison conditions and easily to receive full comparison results, only ARIMA,

W.-C. Hong, *Intelligent Energy Demand Forecasting*, Lecture Notes in Energy 10,
DOI 10.1007/978-1-4471-4968-2_2, © Springer-Verlag London 2013

SARIMA, HW, SHW, GRNN (general regression neural network), BPNN (back-propagation neural network), and SVR models are introduced in the following subsections, whereas the state space/Kalman filtering, linear regression, and KBES models are beyond the scope of this book.

2.1 Autoregressive Integrated Moving Average Model

Introduced by Box and Jenkins [13], the ARIMA model has been one of the most popular approaches in forecasting. The ARIMA model is composed of three partitions: the autoregressive (AR), the moving average (MA), and the differencing process (also called integrated processes). In the AR process, the current value of electric load is often expressed as linear combination of previous actual electric load values and with a random noise. The order of AR process is determined by the oldest previous electric load value that is regressed with the same series itself. In the MA process, it expresses a white noise error series of linear combination in terms of current against previous (unobserved) white noise error term. The order of MA process is determined by the oldest previous value. The AR and MA processes are combined to be the famous electric load forecasting model, autoregressive moving average (ARMA) process. In the ARMA process, the order is determined by the same method as mentioned in AR and MA processes [14].

The AR, MA, or ARMA models are often viewed as stationary processes, that is, their means and covariances are stationary with respect to time. Therefore, while the process is nonstationary, it is necessarily transformed to a stationary series before conducting their modeling processes. Differencing process is employed to transform a nonstationary series into a stationary one. The order of a differencing process is the number of times of differenced before achieving stationarity. Differencing processes for AR, MA, or ARMA models are also the so-called integrated processes and are named as ARI, IMA, and ARIMA, respectively.

In an ARIMA model, the future value of a variable is supposed to be a linear combination of past values and past errors, expressed as Eq. (2.1):

$$y_t = \theta_0 + \phi_1 y_{t-1} + \phi_2 y_{t-2} + \cdots + \phi_p y_{t-p} + \cdots \varepsilon_t - \theta_1 \varepsilon_{t-1} - \theta_2 \varepsilon_{t-2} - \cdots - \theta_q \varepsilon_{t-q}, \tag{2.1}$$

where y_t is the actual value and ε_t is the random error at time t, ϕ_i and θ_j are the coefficients, and p and q are integers and often referred to as autoregressive and moving average polynomials, respectively. In addition, the difference (∇) is used to solve the nonstationary problem and defined as Eq. (2.2):

$$\nabla^d y_t = \nabla^{d-1} y_t - \nabla^{d-1} y_{t-1}. \tag{2.2}$$

Basically, three phases are included in an ARIMA model: model identification, parameter estimation, and diagnostic checking. Furthermore, the backward shift operator, B, is defined as Eqs. (2.3) and (2.4):

$$B^1 y_t = y_{t-1}, B^2 y_t = y_{t-2}, \ldots, B^p y_t = y_{t-p}, \tag{2.3}$$

$$B^1 \varepsilon_t = \varepsilon_{t-1}, B^2 \varepsilon_t = \varepsilon_{t-2}, \ldots, B^p \varepsilon_t = \varepsilon_{t-p}. \tag{2.4}$$

Then $\phi_p(B)$ and $\theta_q(B)$ can be written as Eqs. (2.5) and (2.6), respectively,

$$\phi_p(B) = 1 - \phi_1 B^1 - \phi_2 B^2 - \cdots\cdots - \phi_p B^p, \tag{2.5}$$

$$\theta_1(B) = 1 - \theta_1 B^1 - \theta_2 B^2 - \cdots\cdots - \theta_q B^q. \tag{2.6}$$

Hence, Eq. (2.1) can be rewritten as Eq. (2.7):

$$\phi_p(B)\nabla^d y_t = C_0 + \theta_q(B)\varepsilon_t. \tag{2.7}$$

Equation (2.7) is denoted as ARIMA(p,d,q) with nonzero constant, C_0. For example, the ARIMA(2,2,1) model can be represented as Eq. (2.8):

$$\phi_2(B)\nabla^2 y_t = C_0 + \theta_1(B)\varepsilon_t. \tag{2.8}$$

In general, the values of p, d, and q need to be estimated by autocorrelation function (ACF) and partial autocorrelation function (PACF) of the differenced series.

2.2 Seasonal Autoregressive Integrated Moving Average Model

For a special-period time series, a seasonal or cyclic component should be considered in ARIMA modeling process. This additional process is well known as the seasonal process, and its abbreviation is used as SARIMA [15]. The SARIMA process is often referred to as the SARIMA(p,d,q)(P,D,Q)$_S$ model. Similar to the ARIMA model, the forecasting values are assumed to be a linear combination of past values and past errors. A time series $\{X_t\}$ is a SARIMA process with seasonal period length S if d and D are nonnegative integers and if the differenced series $W_t = (1 - B)^d(1 - B^S)^D X_t$ is a stationary autoregressive moving average process. In symbolic terms, the model can be written as Eq. (2.9):

$$\phi_p(B)\Phi_P(B^S)W_t = \theta_q(B)\Theta_Q(B^S)\varepsilon_t, \quad t = 1, 2, \ldots, N, \tag{2.9}$$

where N is the number of observations up to time t, B is the backshift operator defined by $B^a W_t = W_{t-a}$, $\phi_p(B) = 1 - \phi_1 B - \cdots - \phi_p B^p$ is called a regular (non-seasonal) autoregressive operator of order p, $\Phi_P(B^S) = 1 - \Phi_1 B^S - \cdots - \Phi_P B^{PS}$ is a seasonal autoregressive operator of order P, $\theta_q(B) = 1 - \theta_1 B - \cdots - \theta_q B^q$ is a regular moving average operator of order q, $\Theta_Q(B^S) = 1 - \Theta_1 B^S - \cdots - \Theta_Q B^{QS}$ is a seasonal moving average operator of order Q, and ε_t is identically and independently distributed as normal random variables with mean zero, variance σ^2, and cov $(\varepsilon_t, \varepsilon_{t-k}) = 0, \forall k \neq 0$.

In the definition above, the parameters p and q represent the autoregressive and moving average order, respectively, and the parameters P and Q represent the autoregressive and moving average order at the model's seasonal period length, S, respectively. The parameters d and D represent the order of ordinary and seasonal differencing, respectively.

Basically, when fitting a SARIMA model to data, the first task is to estimate values of d and D, the orders of differencing needed to make the series stationary and to remove most of the seasonality. The values of p, P, q, and Q then need to be estimated by the autocorrelation function (ACF) and partial autocorrelation function (PACF) of the differenced series. Other model parameters may be estimated by suitable iterative procedures.

2.3 Holt–Winters Model

The Holt–Winters (HW) model is proposed by Holt [16] and Winter [17]. HW model is an extension of exponentially weighted moving average procedure. The exponentially weighted moving average approach forecasts future values based on past observations and places more weight on the recent observations. HW model smoothes the trend values separately with two smoothing coefficients (with values between 0 and 1) and incorporates an explicit linear trend in the forecast. The approach of Holt–Winter linear exponential smoothing is shown as Eqs. (2.10)–(2.12):

$$s_t = \alpha a_t + (1 - \alpha)(s_{t-1} + b_{t-1}), \tag{2.10}$$

$$b_t = \beta(s_t - s_{t-1}) + (1 - \beta)b_{t-1}, \tag{2.11}$$

$$f_t = s_t + ib_t, \tag{2.12}$$

where a_t is the actual value at time t, s_t is the smoothed estimate at time t, b_t is the trend value at time t, α is the level smoothing coefficient, and β is the trend smoothing coefficient.

Equation (2.10) lets the actual value be smoothed in a recursive manner by weighting the current level (α) and then adjusts s_t directly for the trend of the previous period, b_{t-1}, by adding it to the last smoothed value, s_{t-1}. This helps to eliminate the lag and brings s_t to the approximate base of the current data value. Equation (2.11) updates the trend, which is expressed as the difference between the last two smoothed values. It modifies the trend by smoothing with β in the last period ($s_t - s_{t-1}$) and adding that to the previous estimate of the trend multiplied by $(1 - \beta)$. Equation (2.12) is used to forecast ahead. The trend, b_t, is multiplied by the number of periods ahead to be forecast, i, and added to the base value, s_t. The forecast error (e_t) is defined as the actual value minus the forecast (fitted) value for time period t, which is shown in Eq. (2.13):

$$e_t = a_t - f_t. \tag{2.13}$$

The forecast error is assumed to be an independent random variable with zero mean and constant variance. Values of smoothing coefficients, α and β, are determined to minimize the forecast error index.

2.4 Seasonal Holt–Winters (SHW) Model

To consider the seasonal effect, the seasonal Holt and Winters' linear exponential smoothing (SHW) approach is also employed. HW model cannot be extended to accommodate additive seasonality if the magnitude of the seasonal effects does not change with the series or multiplicative seasonality if the amplitude of the seasonal pattern changes over time. Therefore, the forecast for SHW model is shown as Eqs. (2.14)–(2.17):

$$s_t = \alpha \frac{a_t}{I_{t-L}} + (1 - \alpha)(s_{t-1} + b_{t-1}), \tag{2.14}$$

$$b_t = \beta(s_t - s_{t-1}) + (1 - \beta)b_{t-1}, \tag{2.15}$$

$$I_t = \gamma \frac{a_t}{s_t} + (1 - \gamma)I_{t-L}, \tag{2.16}$$

$$f_t = (s_t + ib_t)I_{t-L+i}, \tag{2.17}$$

where L is the length of seasonality, I is the seasonal adjustment factor, and γ is the seasonal adjustment coefficient. Equation (2.14) differs slightly from Eq. (2.15) in that the first term is divided by the seasonal number I_{t-L}; this is done to deseasonalize a_t (eliminate seasonal fluctuations from a_t). Equation (2.16) is comparable to a seasonal index that is found as a ratio of current values of the series, a_t, divided by the smoothed value for the series, s_t. If a_t is larger than s_t, the ratio will be greater than 1, else, the ratio will be less than 1. In order to smooth the randomness of a_t, Eq. (2.17) weights the newly computed seasonal factor with γ and the most recent seasonal number corresponding to the same season with $(1-\gamma)$.

2.5 General Regression Neural Network Model

The GRNN model, proposed by Specht [18], can approximate any arbitrary function from historical data. The foundation of GRNN operation is based on the theory of kernel regression. The procedure of the GRNN model can be equivalently represented as Eq. (2.18):

$$E[N|M] = \frac{\int_{-\infty}^{\infty} Nf(M,N)\mathrm{d}N}{\int_{-\infty}^{\infty} f(M,N)\mathrm{d}N}, \tag{2.18}$$

where N is the predicted value of GRNN, M is the input vector (M_1, M_2, ..., M_n) which consists of n variables, $E[N|M]$ is the expected value of the output N given an input vector M, and $f(M, N)$ is the joint probability density function of M and N.

GRNN model primarily has four layers (Fig. 2.1). Each layer is assigned with a specific computational function when nonlinear regression function is performed. The first layer of the network is to receive information. The input neurons then feed the data to the second layer. The primary task of the second layer is to memorize the relationship between the input neuron and its proper response. Therefore, the neurons in the second layer are also called pattern neurons. A multivariate Gaussian function of θ_i is given in Eq. (2.19), and the data from the input neurons are used to compute an output θ_i by a typical pattern neuron i:

$$\theta_i = \exp\left[\frac{-(M - U_i)'(M - U_i)}{2\sigma^2}\right], \tag{2.19}$$

where U_i is a specific training vector represented by pattern neuron i and σ is the smoothing parameter. In the third layer, the neurons, namely, the summation neurons, receive the outputs of the pattern neurons. The outputs from all pattern neurons are augmented. Basically, two summations, the simple summation and the weighted summation, are conducted in the neurons of the third layer. The simple summation and the weighted summation operations can be represented as Eqs. (2.20) and (2.21), respectively:

$$S_{\mathrm{s}} = \sum_i \theta_i, \tag{2.20}$$

$$S_{\mathrm{w}} = \sum_i w_i \theta_i, \tag{2.21}$$

where w_i is the pattern neuron i connected to third layer of weights.

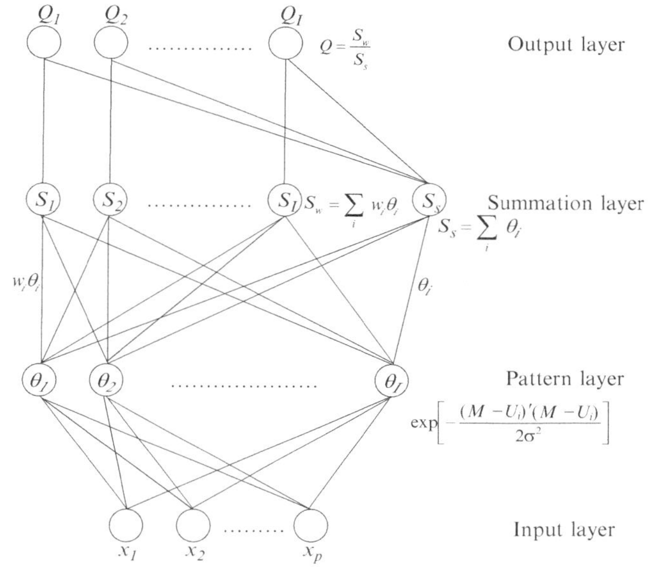

Fig. 2.1 The architecture of the GRNN model

The summations of neurons in the third layer are then fed into the fourth layer. The GRNN regression output Q is calculated as Eq. (2.22):

$$Q = \frac{S_s}{S_w}. \tag{2.22}$$

2.6 Back-Propagation Neural Networks Model

The multilayer back-propagation neural network (BPNN) is one of the most widely used neural network models. Consider the simplest BPNN architecture (Fig. 2.2) including three layers: an input layer (x), an output layer (o), and a hidden layer (h). The computational procedure of this network is described as Eq. (2.23):

$$o_i = f\left(\sum_j g_{ij} x_{ij}\right), \tag{2.23}$$

where o_i denotes the output of node i, $f(\cdot)$ represents the activation function, g_{ij} is the connection weight between nodes i and j in the lower layer which can be replaced with v_{ji} and w_{kj}, and x_{ij} denotes the input signal from the node j in the lower layer.

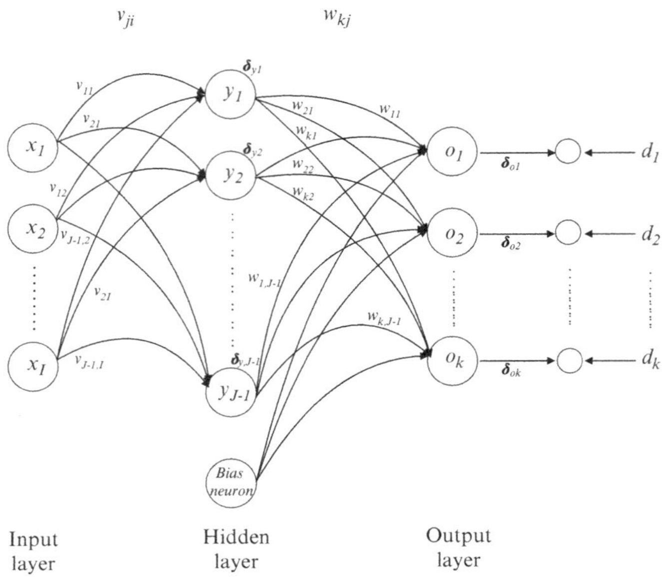

Fig. 2.2 The architecture of the BPNN model

The BPNN algorithm attempts to improve neural network performance (reduce the total error) through changing the gradient weights. The BPNN algorithm minimizes the sum of square error, which can be calculated by Eq. (2.24):

$$E = \frac{1}{2}\sum_{p=1}^{P}\sum_{j=1}^{K}(d_{pj} - o_{pj})^2, \tag{2.24}$$

where E denotes the square errors, K represents the output layer neurons, P is the training data pattern, d_{pj} denotes the actual output, and o_{pj} represents the network output.

The BPNN algorithm is expressed as follows: Let Δv_{jt} denote the weight change for any hidden layer neuron and Δw_{kj} for any output layer neuron, shown as Eqs. (2.25) and (2.26):

$$\Delta v_{ji} = -\eta \frac{\partial E}{\partial v_{ji}} \quad i = 1, \ldots, I, j = 1, \ldots, J-1 \tag{2.25}$$

$$\Delta w_{kj} = -\eta \frac{\partial E}{\partial w_{kj}} \quad j = 1, \ldots, J-1, k = 1, \ldots, K, \tag{2.26}$$

where η represents the learning rate parameter. Notably, the Jth node in Fig. 2.2 is the bias neuron without weight. Equations (2.27) and (2.28) express the signal (s_j) to each hidden layer neuron and the signal (u_k) to each neuron in the output layer:

$$s_j = \sum_{i=1}^{I} v_{ji} x_i, \tag{2.27}$$

$$u_k = \sum_{j=1}^{J-1} w_{kj} y_j. \tag{2.28}$$

The error signal terms for the *j*th hidden neuron δ_{yj} and for the *k*th output neuron δ_{ok} are defined as Eqs. (2.29) and (2.30), respectively:

$$\delta_{yj} = -\frac{\partial E}{\partial s_j}, \tag{2.29}$$

$$\delta_{ok} = -\frac{\partial E}{\partial u_k}. \tag{2.30}$$

Applying the chain rule, the gradient of the cost function with respect to weights v_{ji} and w_{kj} is

$$\frac{\partial E}{\partial v_{ji}} = \frac{\partial E}{\partial s_j}\frac{\partial s_j}{\partial v_{ji}}, \tag{2.31}$$

$$\frac{\partial E}{\partial w_{kj}} = \frac{\partial E}{\partial u_k}\frac{\partial u_k}{\partial w_{kj}}, \tag{2.32}$$

and

$$\frac{\partial s_j}{\partial v_{ji}} = \frac{\partial(v_{j1}x_1 + v_{j2}x_2 + \cdots + v_{ji}x_i + \cdots + v_{jI}x_I)}{\partial v_{ji}} = x_i, \tag{2.33}$$

$$\frac{\partial u_k}{\partial w_{kj}} = \frac{\partial(w_{k1}y_1 + w_{k2}y_2 + \cdots + w_{kj}y_j + \cdots + w_{kJ}y_J)}{\partial w_{kj}} = y_j. \tag{2.34}$$

By combining Eqs. (2.29), (2.31), and (2.33) and Eqs. (2.30),(2.32), and (2.34), one obtains Eqs. (2.35) and (2.36):

$$\frac{\partial E}{\partial v_{ji}} = -\delta_{yj} x_i, \tag{2.35}$$

$$\frac{\partial E}{\partial w_{kj}} = -\delta_{ok} y_j. \tag{2.36}$$

The weight change from Eqs. (2.25) and (2.26) can now be written as Eqs. (2.37) and (2.38), respectively:

$$\Delta v_{ji} = -\eta \frac{\partial E}{\partial v_{ji}} = \eta \delta_{yj} x_i, \tag{2.37}$$

$$\Delta w_{kj} = -\eta \frac{\partial E}{\partial e_{kj}} = \eta \delta_{ok} y_j. \tag{2.38}$$

Furthermore, Eqs. (2.29) and (2.30) can be calculated as Eqs. (2.39) and (2.40):

$$\delta_{ok} = -\frac{\partial E}{\partial u_k} = -\frac{\partial E}{\partial o_k}\frac{\partial o_k}{\partial u_k} = (d_k - o_k) f'(u_k), \tag{2.39}$$

$$\delta_{yj} = -\frac{\partial E}{\partial s_j} = -\frac{\partial E}{\partial y_j}\frac{\partial y_j}{\partial s_j} = \left\{ \sum_{k=1}^{K} o_k \cdot w_{kj} \right\} \cdot f'_j(u_j). \tag{2.40}$$

The weights, v_{ji} and w_{kj}, are changed as Eqs. (2.41) and (2.42):

$$w_{kj} = w_{kj} + \Delta w_{kj} = w_{kj} + \eta \delta_{ok} y_j, \tag{2.41}$$

$$v_{ji} = v_{ji} + \Delta v_{ji} = v_{ji} + \eta f'_j(u_j) x_i \sum_{k=1}^{K} \delta_{ok} w_{kj}. \tag{2.42}$$

The constant term, η, is specified at the start of training cycle and determines the training speed and stability of the network. The most common activation functions are the squashing sigmoid function, such as the logistic and tangent hyperbolic functions.

2.7 Support Vector Regression Model

2.7.1 Structural Risk Minimization

Artificial intelligent approaches have tended to be based on finding functions to map as training errors over training set, that is, empirical risk minimization (ERM). However, the ERM does not guarantee good generalization to novel testing data set. To separate the classes with a surface (hyperplane) that maximizes the margin between training data set, SVMs employ the SRM principle that aims to minimize a bound on the generalization error, rather than minimizing the mean square error over the training data set. SRM provides a well-defined quantitative measurement for the capacity of a learned function to capture the true structure of the data distribution and generalize over unknown test data set. Vapnik–Chervonenkis (VC) dimension [19] has been applied for such a capacity; by selecting a function and minimizing its empirical error to a training data set, SRM can guarantee a minimal bound on the test data set.

Give a training data set of N elements $\{(\mathbf{x}_i, y_i), i = 1, 2, \ldots, N\}$, where $\mathbf{x}_i$ is the ith element in n-dimensional space, that is, $\mathbf{x}_i = [x_{1i}, \ldots, x_{ni}] \in \Re^n$, and $y_i \in \{-1, +1\}$ is the label of $\mathbf{x}_i$, then define a deterministic function f:$\mathbf{x} \rightarrow \{-1,+1\}$ for a given input data $\mathbf{x}$ and adjustable weights $\mathbf{w}(\mathbf{w} \in \Re^n)$, according to the same but unknown probability distribution ($P(\mathbf{x}, y)$). The weights $\mathbf{w}$ would be adjusted during the training stage. Since the underlying probability distribution $P(\mathbf{x}, y)$ is unknown, the upper bound for the probability of classification errors on the test data set (i.e., expected error of f, $R(f)$) cannot be minimized directly. Thus, it is feasible to estimate an approximate function of $R(f)$, that is, empirical risk, denoted as $R_{\text{emp}}(f)$, that is close to the optimal one based on the training data pairs ($\mathbf{x}$, y). Then, according to the SRM principle [20, 21], $R(f)$ and $R_{\text{emp}}(f)$ are expressed as Eqs. (2.43)–(2.46):

$$R(f) \leq R_{\text{emp}}(f) + \varepsilon_1(N, h, \eta, R_{\text{emp}}), \tag{2.43}$$

$$R_{\text{emp}}(f) = \frac{1}{N}\sum_{i=1}^{N} |y_i - f(\mathbf{x}_i)|_{\text{loss function.}}, \tag{2.44}$$

$$\varepsilon_1(N, h, \eta, R_{\text{emp}}) = 2\varepsilon_0^2(N, h, \eta)\left(1 + \sqrt{1 + \frac{R_{\text{emp}}(f)}{\varepsilon_0^2(N, h, \eta)}}\right), \tag{2.45}$$

$$\varepsilon_0(N, h, \eta) = \sqrt{\frac{h\left(\ln\left(\frac{2N}{h}\right) + 1\right) - \ln\left(\frac{\eta}{4}\right)}{N}}. \tag{2.46}$$

Equation (2.43) holds with probability $1-\eta$ for $0 \leq \eta \leq 1$. The $\varepsilon_0(N,h,\eta)$ is the so-called VC confidence interval. The values of $\varepsilon_0(N,h,\eta)$ depend on the number of training data N, the VC dimension h, and the value of η.

For a small empirical risk $R_{\text{emp}}(f)$, for example, closes to 0, then Eq. (2.43) would approximately reduce to $R_{\text{emp}}(f) + 4{\varepsilon_0}^2(N,h,\eta)$ and, in contrast, for a large empirical risk closes to 1, Eq. (2.43) would approximately reduce to $R_{\text{emp}}(f) + \varepsilon_0(N, h,\eta)$ [22].

Thus, there are two strategies for minimizing the upper bound, $R(f)$. The first one is to keep the VC confidence ($\varepsilon_0(N,h,\eta)$) fixed and to minimize the empirical risk; most of ANN models seek to employ the first one. However, the first way does not perform well because dealing with $R_{\text{emp}}(f)$ lonely cannot guarantee reduction of VC confidence. The second one is to fix the empirical risk to a small value and to minimize the VC confidence, which is the so-called SRM principle. Although

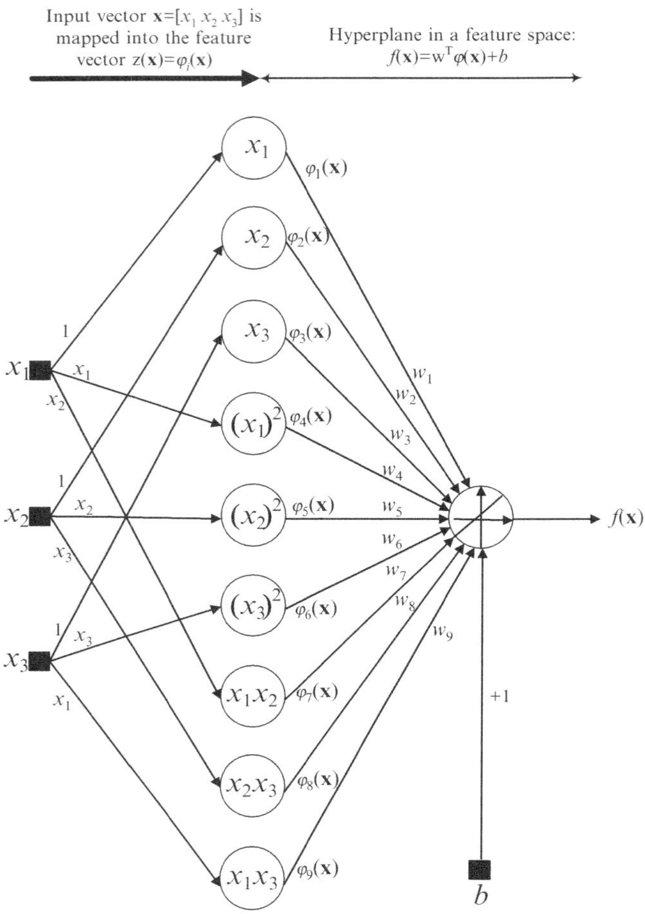

Fig. 2.3 Transformation of the second-order polynomial hyperplane over a three-dimensional original space in an SVR model

SVMs implement this principle, their training algorithm that aims to minimize the VC dimension is still based on a hierarchy that depends on the data [20, 23].

2.7.2 *Support Vector Regression*

As mentioned above, SVMs have originally been used for classification purposes, but their principles can be extended easily to the task of regression and time series prediction. The brief ideas of SVMs for the case of regression are introduced. A nonlinear mapping $\varphi(\cdot) : \Re^n \rightarrow \Re^{n_h}$ is defined to map the input data (training data

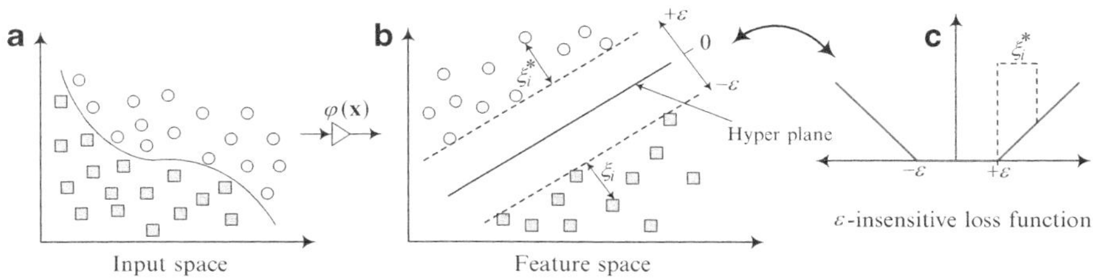

Fig. 2.4 Transformation process illustration of an SVR model

set) $\{(\mathbf{x}_i, y_i)\}_{i=1}^{N}$ into a so-called high-dimensional feature space (Fig. 2.3), which may have infinite dimensions, $\Re^{n_h}$. Then, in the high-dimensional feature space, there theoretically exists a linear function, f, to formulate the nonlinear relationship between input data and output data (Fig. 2.4a, b). Such a linear function, namely, SVR function, is as Eq. (2.47):

$$f(\mathbf{x}) = \mathbf{w}^T \varphi(\mathbf{x}) + b, \tag{2.47}$$

where $f(\mathbf{x})$ denotes the forecasting values and the coefficients $\mathbf{w}$ ($\mathbf{w} \in \Re^{n_h}$) and b ($b \in \Re$) are adjustable. As mentioned above, using SVM method one aims at minimizing the empirical risk as Eq. (2.48):

$$R_{\text{emp}}(f) = \frac{1}{N} \sum_{i=1}^{N} \Theta_\varepsilon(y_i, \mathbf{w}^T \varphi(\mathbf{x}_i) + b), \tag{2.48}$$

where $\Theta_\varepsilon(\mathbf{y}, f(\mathbf{x}))$ is the ε-insensitive loss function (as thick line in Fig. 2.4c) and is defined as Eq. (2.49):

$$\Theta_\varepsilon(\mathbf{y}, f(\mathbf{x})) = \begin{cases} |f(\mathbf{x}) - \mathbf{y}| - \varepsilon, & \text{if } |f(\mathbf{x}) - \mathbf{y}| \geq \varepsilon \\ 0, & \text{otherwise} \end{cases}. \tag{2.49}$$

In addition, $\Theta_\varepsilon(\mathbf{y}, f(\mathbf{x}))$ is employed to find out an optimum hyperplane on the high-dimensional feature space (Fig. 2.4b) to maximize the distance separating the training data into two subsets. Thus, the SVR focuses on finding the optimum hyperplane and minimizing the training error between the training data and the ε-insensitive loss function.

Then, the SVR minimizes the overall errors, shown as Eq. (2.50):

$$\underset{\mathbf{w}, b, \xi^*, \xi}{\text{Min}} \; R_\varepsilon(\mathbf{w}, \xi^*, \xi) = \frac{1}{2}\mathbf{w}^T\mathbf{w} + C\sum_{i=1}^{N} (\xi_i^* + \xi_i), \tag{2.50}$$

with the constraints

$$\begin{aligned}
&\mathbf{y}_i - \mathbf{w}^T\varphi(\mathbf{x}_i) - b \leq \varepsilon + \xi_i^*, \quad i = 1, 2, \ldots, N \\
&-\mathbf{y}_i + \mathbf{w}^T\varphi(\mathbf{x}_i) + b \leq \varepsilon + \xi_i, \quad i = 1, 2, \ldots, N \\
&\xi_i^* \geq 0, \quad i = 1, 2, \ldots, N \\
&\xi_i \geq 0, \quad i = 1, 2, \ldots, N
\end{aligned}.$$

The first term of Eq. (2.50), employing the concept of maximizing the distance of two separated training data, is used to regularize weight sizes, to penalize large weights, and to maintain regression function flatness. The second term penalizes training errors of $f(\mathbf{x})$ and $\mathbf{y}$ by using the ε-insensitive loss function. C is a parameter to trade off these two terms. Training errors above $+\varepsilon$ are denoted as ξ_i^*, whereas training errors below $-\varepsilon$ are denoted as ξ_i (Fig. 2.4b).

After the quadratic optimization problem with inequality constraints is solved, the parameter vector $\mathbf{w}$ in Eq. (2.51) is obtained:

$$\mathbf{w} = \sum_{i=1}^{N} (\beta_i^* - \beta_i)\varphi(\mathbf{x}_i), \tag{2.51}$$

where β_i^*, β_i are obtained by solving a quadratic program and are the Lagrangian multipliers. Finally, the SVR regression function is obtained as Eq. (2.52) in the dual space:

$$f(\mathbf{x}) = \sum_{i=1}^{N} (\beta_i^* - \beta_i)K(\mathbf{x}_i, \mathbf{x}) + b, \tag{2.52}$$

where $K(\mathbf{x}_i, \mathbf{x}_j)$ is called the kernel function and the value of the kernel equals the inner product of two vectors, $\mathbf{x}_i$ and $\mathbf{x}_j$, in the feature space $\varphi(\mathbf{x}_i)$ and $\varphi(\mathbf{x}_j)$, respectively; that is, $K(\mathbf{x}_i, \mathbf{x}_j) = \varphi(\mathbf{x}_i)\circ\varphi(\mathbf{x}_j)$. Any function that meets Mercer's condition [20] can be used as the kernel function.

There are several types of kernel function. The most used kernel functions are the Gaussian RBF with a width of σ: $K(\mathbf{x}_i, \mathbf{x}_j) = \exp(-0.5\|\mathbf{x}_i - \mathbf{x}_j\|^2/\sigma^2)$, the polynomial kernel with an order of d and constants a_1 and a_2: $K(\mathbf{x}_i, \mathbf{x}_j) = (a_1 \mathbf{x}_i\mathbf{x}_j + a_2)^d$, and $K(\mathbf{x}_i, \mathbf{x}_j) = \tanh(\mathbf{x}_i^T\mathbf{x}_j - b)$, where b is a constant, in which, if the value of σ is very large, the RBF kernel approximates the use of a linear kernel (polynomial with an order of 1). Till now, it is hard to determine the type of kernel functions for specific data patterns [24, 25]. However, based on Smola et al.'s [26] empirical results, they claim that the Gaussian RBF kernel is not only easier to implement but also capable to nonlinearly map the training data into an infinite-dimensional space; thus, it is suitable to deal with nonlinear relationship problems. Therefore, the Gaussian RBF kernel function is specified in this book. The forecasting process of an SVR model is illustrated as in Fig. 2.5.

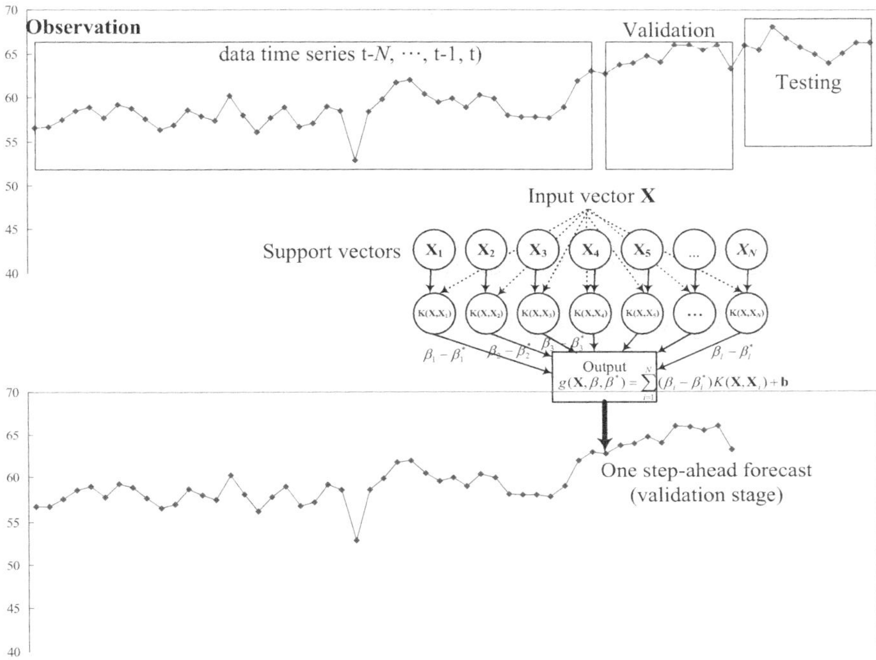

Fig. 2.5 The forecasting process of an SVR model

2.7.3 *The Role of Evolutionary Algorithms*

It is well known that the forecasting accuracy of an SVR model depends on a good setting of hyper parameters C, ε, and the kernel parameters (σ). For example, parameter C is considered to specify the trade-off between the model flatness and the degree of the training errors larger than ε which are tolerated in Eq. (2.48) (i.e., the empirical risk). If C is too large (approximated to infinity), then the objective is only to minimize the empirical risk, $\Theta_\varepsilon(\mathbf{y}, f(\mathbf{x}))$, without model flatness in the optimization formulation, Eq. (2.48). Parameter ε controls the width of the ε-insensitive loss function, that is, the number of support vectors (SVs) employed by the regression [20]. Larger ε-value results in fewer SVs employed; thus, the regression function is more flat (simple). Parameter σ controls the Gaussian function width, which reflects the distribution range of $\mathbf{x}$-values of training data. Therefore, the three parameters affect model constructing in different ways. There is no structural method or any shortage opinions on efficient setting of SVR parameters. Although numerous publications in the literature have given some recommendations on appropriate setting of SVR parameters [27], however, those approaches do not simultaneously consider the interaction effects among the three parameters. Thus, the determination of these three parameters' selection is further an important issue.

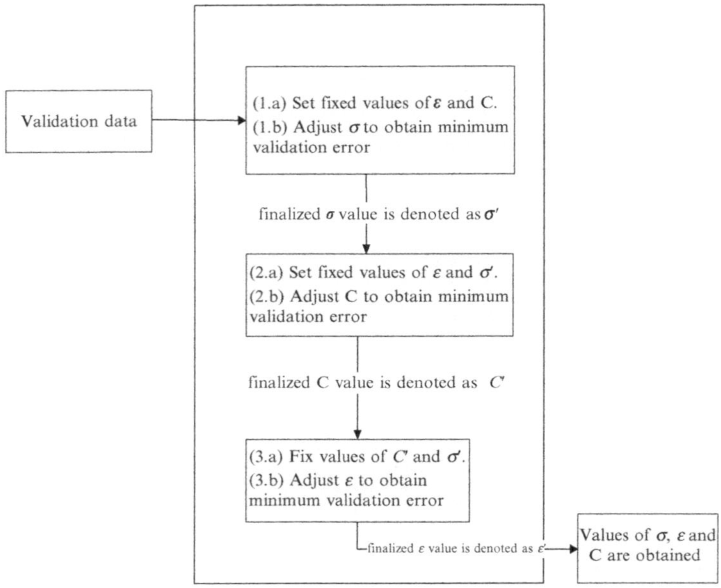

Fig. 2.6 The traditional determining processes of three parameters in an SVR model

The traditional determination procedure in determining suitable values of these three parameters is conducted in following steps and shown as Fig. 2.6:

Step 1: Set fixed values of the parameters ε and C. Then, adjust the value of σ till a minimum testing error is achieved. The finalized σ value is denoted as σ'.

Step 2: Set a fixed value of the parameter ε, and the value of σ is set to σ'. Then, adjust the value of C to achieve a minimum testing error. The finalized C is defined as C'.

Step 3: Values of σ and C are set to σ' and C'. Then, adjust ε till a minimum testing error is obtained. The finalized ε is defined as ε'. Therefore, values of σ, ε, and C are obtained as σ', ε', and C'.

Some numerical examples are used to illustrate step by step the traditional determination procedure. For the first example, please refer Figs. 2.7, 2.8, and 2.9. First, the fixed values, 0.5 and 50, are set to the parameters σ and C, respectively. Then, adjust the value of ε till a minimum testing error (NRMSE) is reached. It is found that there exists only one local minimum value of NRMSE (0.008049) when $\varepsilon = 0.27$, shown in Fig. 2.7. Therefore, set finalized ε-value as 0.27. Second, set $\varepsilon = 0.27$ and $C = 50$, respectively. It is found that there also exists only one local minimum value of NRMSE (0.007584) when $\sigma = 0.45$, presented in Fig. 2.8. Finally, the third step will focus on searching the third parameter, set to $\sigma = 0.45$ and $\varepsilon = 0.27$; it is found that there exists only one local minimum value of NRMSE (0.006446) when $C = 60$, presented in Fig. 2.9. Therefore, the three parameters are estimated as follows: $\sigma = 0.45$, $C = 60$, and $\varepsilon = 0.27$.

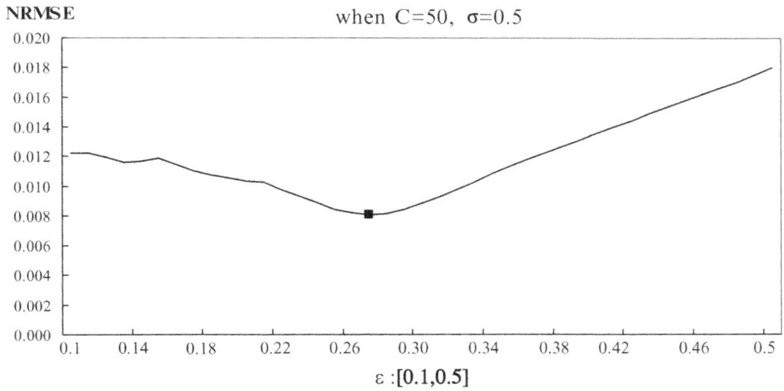

Fig. 2.7 Fixed values of C and σ to adjust ε

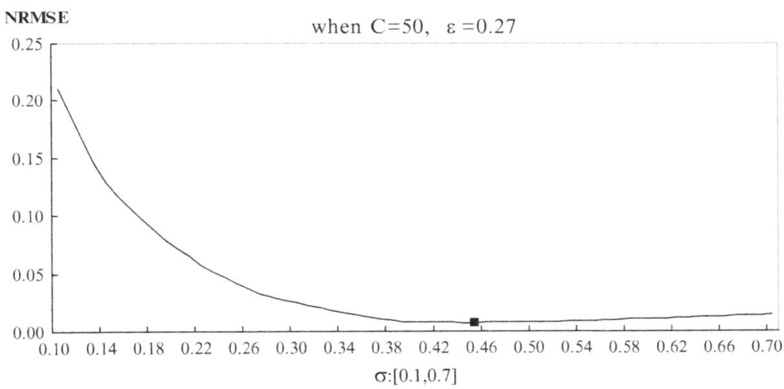

Fig. 2.8 Fixed values of C and ε to adjust σ

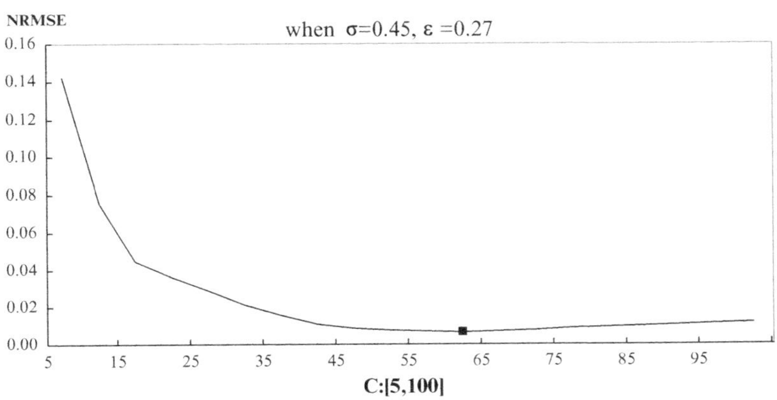

Fig. 2.9 Fixed values of σ and ε to adjust C

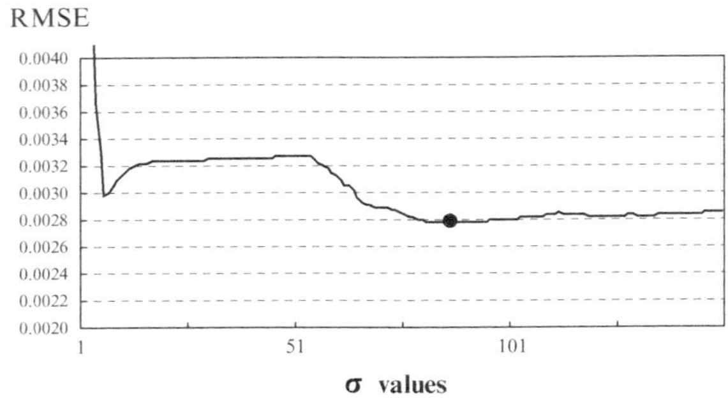

Fig. 2.10 Fixed values of C and ε to adjust σ values

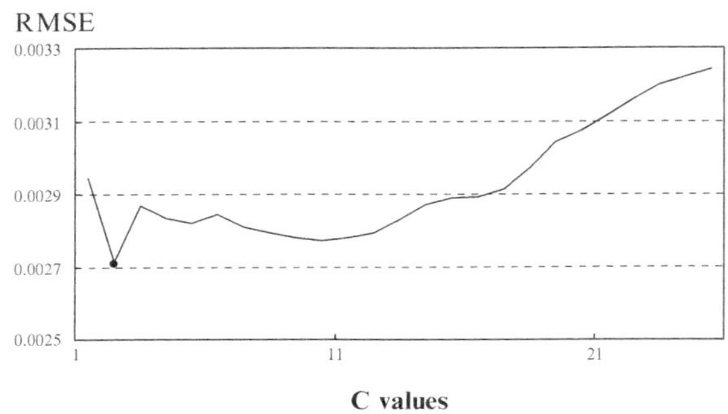

Fig. 2.11 Fixed values of σ and ε to adjust C values

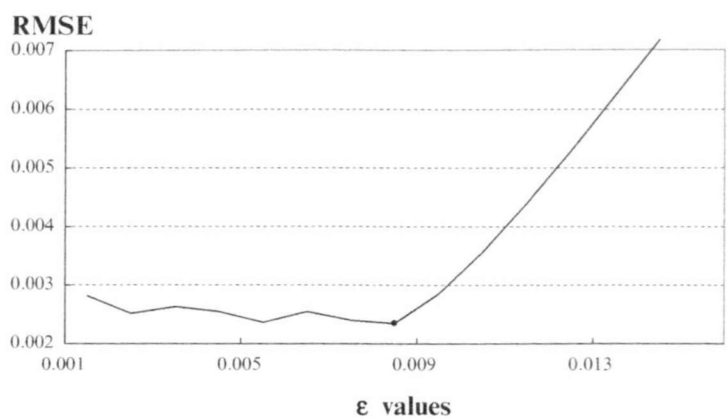

Fig. 2.12 Fixed values of σ and C to adjust ε-values

For the second example, please refer Figs. 2.10, 2.11, and 2.12. First, set the values of the parameters ε and C fixed at 0 and 10, respectively. Then, adjust the value of σ till a minimum testing error (RMSE) is reached. In Fig. 2.10, when the finalized σ value equaled to 87, the minimum value of RMSE is achieved. Therefore, set finalized σ value as 87. Secondly, fix the values of the parameters σ and ε at 87 and 0, respectively. Then, adjust the value of C (Fig. 2.11). The finalized value of C is 2. The minimum value of RMSE is obtained. Therefore, the finalized C value is 2. Finally, set the values of the parameters σ and C fixed at

87 and 2, respectively. Then, adjust the value of ε (Fig. 2.12). The ε-value equals to 0.008 when the minimum value of RMSE is achieved. Therefore, the values of three parameters (σ, C, and ε) are 87, 2, and 0.008, correspondingly. The minimum value of RMSE is 0.002345.

The traditional determination of these three parameters is not only time-consuming but also unable to receive satisfied forecasting accuracy level. This is because it is difficult to set up more suitable initial values of parameters ε and C in the initial step and it cannot efficiently find out the near-optimal solution for large-scale data set; particularly, while simultaneously considering the interaction effects among the three parameters, the computing complexity will exhaust the limited decision time. Therefore, it is feasible to employ optimization solving procedure to obtain suitable parameter combination, such as minimizing the objective function describing the structural risk mentioned above. Evolutionary algorithms, such as genetic algorithm, simulated annealing algorithms, immune algorithms, particle swarm optimization, and tabu search, are the very candidates to be employed to determine appropriate parameter values. The author will start a series exploration by employing different evolutionary algorithms to determine suitable parameters of an SVR model in Chap. 3.

References

1. Moghram I, Rahman S (1989) Analysis and evaluation of five short-term load forecasting techniques. IEEE Trans Power Syst 4:1484–1491. doi:10.1109/59. 41700
2. El-Hawary ME, Mbamalu GAN (1990) Short-term power system load forecasting using the iteratively reweighted least squares algorithm. Electric Power Syst Res 19:11–22. doi:10.1016/0378-7796(90)90003-L
3. Papalexopoulos AD, Hesterberg TC (1990) A regression-based approach to short-term system load forecasting. IEEE Trans Power Syst 5:1535–1547. doi:10.1109/59.99410
4. Grady WM, Groce LA, Huebner TM, Lu QC, Crawford MM (1991) Enhancement, implementation and performance of an adaptive short-term load forecasting algorithm. IEEE Trans Power Syst 6:1404–1410. doi:10.1109/59. 116982
5. Al-Fuhaid AS, EL-Sayed MA, Mahmoud MS (1997) Cascaded artificial neural networks for short-term load forecasting. IEEE Trans Power Syst 12:1524–1529. doi:10.1109/59.627852
6. Hong WC (2009) Hybrid evolutionary algorithms in a SVR-based electric load forecasting model. Int J Electr Power Energ Syst 31:409–417. doi:10.1016/j.ijepes.2009.03.020
7. Hong WC (2009) Chaotic particle swarm optimization algorithm in a support vector regression electric load forecasting model. Energy Convers Manage 50:105–117. doi:10.1016/j.enconman.2008.08.031
8. Hong WC (2009) Electric load forecasting by support vector model. Appl Math Model 33:2444–2454. doi:10.1016/j.apm.2008.07.010
9. Hong WC (2010) Application of chaotic ant swarm optimization in electric load forecasting. Energ Policy 38:5830–5839. doi:10.1016/j.enpol.2010.05.033
10. Pai PF, Hong WC (2005) Forecasting regional electric load based on recurrent support vector machines with genetic algorithms. Electric Power Syst Res 74:417–425. doi:10.1016/j.epsr.2005.01.006

11. Pai PF, Hong WC (2005) Support vector machines with simulated annealing algorithms in electricity load forecasting. Energy Convers Manage 46:2669–2688. doi:10.1016/j.enconman.2005.02.004
12. Hong WC, Dong Y, Zhang WY, Chen LY, Panigrahi BK (2013) Cyclic electric load forecasting by seasonal SVR with chaotic genetic algorithm. Int J Electr Power Energ Syst 44:604–614. doi:10.1016/j.ijepes.2012.08.010
13. Box GEP, Jenkins GM (1970) Time series analysis, forecasting and control. Holden-Day, San Francisco, CA
14. Abu-El-Magd MA, Sinha NK (1982) Short-term load demand modeling and forecasting: a review. IEEE Trans Syst Man Cybern 12:370–382. doi:10.1109/TSMC.1982.4308827
15. Soliman SA, Persaud S, El-Nagar K, El-Hawary ME (1997) Application of least absolute value parameter estimation based on linear programming to short-term load forecasting. Int J Electr Power Energ Syst 19:209–216. doi:10.1016/S0142-0615(96)00048-8
16. Holt CC (1957) Forecasting seasonal and trends by exponentially weighted averages. Carnegie Institute of Technology, Pittsburgh, PA
17. Winters PR (1960) Forecasting sales by exponentially weighted moving averages. Manag Sci 6:324–342. doi:10.1287/mnsc.6.3.324
18. Specht DA (1991) A general regression neural network. IEEE Trans Neural Netw 2:568–576. doi:10.1109/72.97934
19. Cao L, Gu Q (2002) Dynamic support vector machines for non-stationary time series forecasting. Intell Data Anal 6:67–83
20. Vapnik V (1995) The nature of statistical learning theory. Springer, New York, NY
21. Cortes C, Vapnik V (1995) Support vector networks. Mach Learn 20:273–297. doi:10.1023/A:1022627411411
22. Haykin S (1999) Neural networks: a comprehensive foundation. Prentice-Hall, Upper Saddle River, NJ
23. Shawe-Taylor J, Bartlett PL, Williamson RC, Anthony M (1998) Structural risk minimization over data-dependent hierarchies. IEEE Trans Inf Theory 44:1926–1940. doi:10.1109/18.705570
24. Amari S, Wu S (1999) Improving support vector machine classifiers by modifying kernel functions. Neural Netw 12:783–789. doi:10.1016/S0893- 6080(99)00032-5
25. Vojislav K (2001) Learning and soft computing—support vector machines, neural networks and fuzzy logic models. The MIT Press, Cambridge, MA, 2001
26. Smola AJ, Schölkopf B, Müller KR (1998) The connection between regularization operators and support vector kernels. Neural Netw 11:637–649. doi:10.1016/S0893-6080(98)00032-X
27. Cherkassky V, Ma Y (2004) Practical selection of SVM parameters and noise estimation for SVM regression. Neural Netw 17:113–126. doi:10.1016/S0893-6080(03)00169-2

Chapter 3
Evolutionary Algorithms in SVR's Parameter Determination

As mentio.ned in Chap. 2, the traditional determination of three parameters does not guarantee to improve forecasting accuracy level, because of its inability to set up more suitable initial values of parameters σ, C, and ε in the initial step and inability to simultaneously consider the interaction effects among three parameters to efficiently find out the near optimal solution for large scale data set. Therefore, it is feasible to employ evolutionary algorithms to conduct intelligent searching around the solution range to determine suitable parameter combination by minimizing the objective function describing the structural risk of a SVR model. This chapter introduces several representative evolutionary algorithms used in a SVR forecasting model to determine suitable parameter combination to receive improved forecasting accuracy level.

3.1 Data Set and Forecasting Comparison Statistical Tests

3.1.1 Data Set

To be based on the same comparison conditions, this book uses historical monthly electric load data of Northeast China to compare the forecasting performance among the proposed SVR-based models hybridizing with seven evolutionary algorithms, ARIMA, and TF-ε-SVR-SA models proposed by Wang et al. [1]. Table 3.1 lists the data set used in this chapter. Figure 3.1 illustrates the tendency of this data set. Totally, there are 64 data (from January 2004 to April 2009) of Northeastern China monthly electric load. However, based on Wang et al.'s [1] support vectors computation, only 53 months data (from December 2004 to April 2009) are suggested. Therefore, for the same comparison condition, the employed data are divided into three data sets: the training data set (32 months, December

W.-C. Hong, *Intelligent Energy Demand Forecasting*, Lecture Notes in Energy 10,
DOI 10.1007/978-1-4471-4968-2_3,

Table 3.1 Monthly electric load in Northeastern China (from Jan. 2004 to Apr. 2009) (unit, hundred million kWh)

Time	Electric load	Time	Electric load	Time	Electric load
Jan. 2004	129.08	Nov. 2005	150.84	Sep. 2007	175.41
Feb. 2004	127.24	Dec. 2005	165.27	Oct. 2007	179.64
Mar. 2004	136.95	Jan. 2006	155.31	Nov. 2007	188.89
Apr. 2004	125.34	Feb. 2006	138.5	Dec. 2007	197.62
May 2004	126.86	Mar. 2006	133.27	Jan. 2008	200.35
Jun. 2004	129.34	Apr. 2006	151.41	Feb. 2008	169.24
Jul. 2004	131.91	May 2006	155.63	Mar. 2008	196.97
Aug. 2004	136.22	Jun. 2006	155.7	Apr. 2008	186.15
Sep. 2004	131.56	Jul. 2006	162.98	May 2008	188.485
Oct. 2004	134.62	Aug. 2006	163.41	Jun. 2008	190.82
Nov. 2004	144.62	Sep. 2006	157.57	Jul. 2008	196.53
Dec. 2004	154.62	Oct. 2006	160.15	Aug. 2008	197.67
Jan. 2005	151.48	Nov. 2006	168.13	Sep. 2008	183.77
Feb. 2005	126.74	Dec. 2006	180.71	Oct. 2008	181.07
Mar. 2005	148.57	Jan. 2007	179.94	Nov. 2008	180.56
Apr. 2005	136.6	Feb. 2007	147.29	Dec. 2008	189.03
May 2005	138.83	Mar. 2007	172.45	Jan. 2009	182.07
Jun. 2005	136.6	Apr. 2007	169.98	Feb. 2009	167.35
Jul. 2005	146.21	May 2007	173.21	Mar. 2009	189.3
Aug. 2005	146.09	Jun. 2007	177.43	Apr. 2009	175.84
Sep. 2005	140.04	Jul. 2007	184.29		
Oct. 2005	142.02	Aug. 2007	183.53		

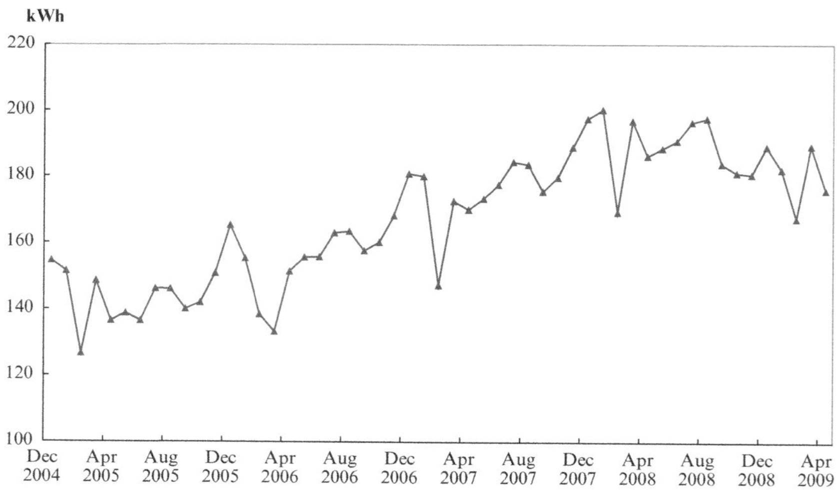

Fig. 3.1 Illustration of the monthly electric load in Northeastern China (from Jan. 2004 to Apr. 2009)

Table 3.2 Training, validation, and testing data sets of the proposed models

Data sets	SVR-based models	TF-ε-SVR-SA model [1]
Training data	Dec. 2004–Jul. 2007	Dec. 2004–Sep. 2008
Validation data	Aug. 2007–Sep. 2008	
Testing data	Oct. 2008–Apr. 2009	Oct. 2008–Apr. 2009

2004 to July 2007), the validation data set (14 months, August 2007 to September 2008), and the testing data set (7 months, from October 2008 to April 2009), as shown in Table 3.2.

3.1.2 Forecasting Comparison Statistical Tests

3.1.2.1 Wilcoxon Signed-Rank Test

Wilcoxon signed-rank test is used to detect the significance of a difference in central tendency of two data series when the size of two data series is equal [2]. The statistic W is represented as Eq. (3.1):

$$W = \min\{S^+, S^-\}, \tag{3.1}$$

where

$$S^+ = \sum_{i=1}^{n} I^+(d_i), \tag{3.2}$$

$$S^- = \sum_{i=1}^{n} I^-(d_i), \tag{3.3}$$

$$I^+(d_i) = \begin{cases} 1 & \text{if } d_i > 0 \\ 0 & \text{otherwise} \end{cases}, \tag{3.4}$$

$$I^-(d_i) = \begin{cases} 1 & \text{if } d_i < 0 \\ 0 & \text{otherwise} \end{cases}, \tag{3.5}$$

and

$$d_i = (\text{data series I})_i - (\text{data series II})_i. \tag{3.6}$$

3.1.2.2 Asymptotic Test

Diebold and Mariano [3] suggest the asymptotic test (S_1), which is applicable because one or more assumptions are violated to be suitable for the simple F test, the Morgan–Granger–Newbold test [4, 5], and the Meese–Rogoff test [6]. Therefore, the S_1 statistic, as Eq. (3.7), is employed in this study:

$$S_1 = \frac{\bar{d}}{\sqrt{\frac{2\pi \hat{f}_d(0)}{T}}}, \tag{3.7}$$

where d_i is the loss-differential series of two compared electric forecasting models, as shown in Eq. (3.8),

$$d_i = e_{1i}^2 - e_{2i}^2, \tag{3.8}$$

where e_1 and e_2 denoted the errors of the two compared electric load forecasting models, respectively.

$2\pi \hat{f}_d(0)$ is the weighted sum of the available sample autocovariances as Eq. (3.9):

$$2\pi \hat{f}_d(0) = \sum_{\tau=-(T-1)}^{T-1} 1 * \left(\frac{\tau}{S(T)}\right) \hat{\gamma}_d(T), \tag{3.9}$$

where T is the sample size, $\hat{\gamma}_d(T)$ is defined as Eq. (3.10),

$$\hat{\gamma}_d(T) = \frac{1}{T} \sum_{t=|\tau|+1}^{T} (d_t - \bar{d})(d_{t-|\tau|} - \bar{d}), \tag{3.10}$$

and $1 * \left(\frac{\tau}{S(T)}\right)$ is the lag window, defined as Eq. (3.11),

$$1 * \left(\frac{\tau}{S(T)}\right) = \begin{cases} 1 & \text{if } \left|\frac{\tau}{S(T)}\right| \leq 1 \\ 0 & \text{otherwise} \end{cases}. \tag{3.11}$$

Obviously, $S(T) = k$, where k denotes as the number of forecasting ahead, the forecasts in this book are one-step-ahead; thus, k is set as 1.

The test will be performed at the 0.05 and 0.10 significance levels in two-tail tests (with normal distribution) under the null hypothesis of equal forecast accuracy for the two compared electric load forecasting models.

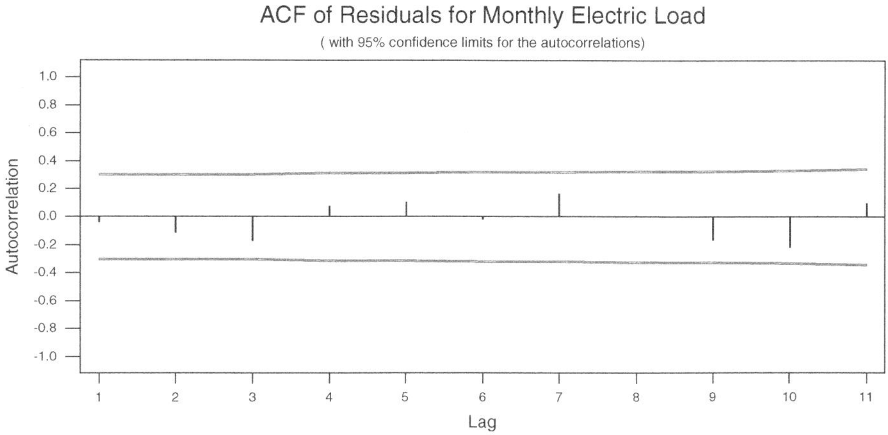

Fig. 3.2 Estimated residual ACF

3.2 Modeling and Forecasting Results of Alternative Models

3.2.1 ARIMA Model

For ARIMA models, the statistical package identified the most suitable model for the training data as ARIMA(1,1,1) with constant term. The ARIMA(1,1,1) model can be expressed as Eq. (3.12):

$$(1 - 0.0641B^1)\nabla y_t = 1.2652 + (1 + 0.9715B^1)\varepsilon_t. \tag{3.12}$$

After determining the suitable parameters of the ARIMA model, it is important to examine how closely the proposed model fits a given time series. The autocorrelation function (ACF) is calculated to verify the parameters. Figure 3.2 plots the estimated residual ACF and indicates that the residuals are not autocorrelated. PACF, the partial autocorrelation function, displayed in Fig. 3.3, is also used to check the residuals and indicates that the residuals are not correlated. The forecasting results are shown in the third column of Table 3.3.

3.2.2 Holt–Winters Model

For the Holt–Winters (HW) method, by Minitab 14 statistic software, the α value and β value are determined as 0.5618 and 0.0472, respectively. The forecasting results are shown in the fourth column of Table 3.3.

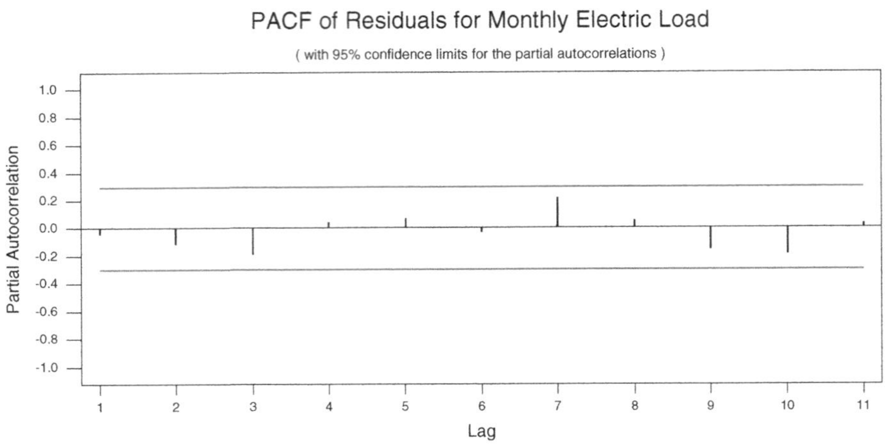

Fig. 3.3 Estimated residual PACF

Table 3.3 Forecasting results of ARIMA, HW, GRNN, and BPNN models (unit, hundred million kWh)

Time point (month)	Actual	ARIMA (1,1,1)	HW (0.5618, 0.0472)	GRNN ($\sigma = 3.33$)	BPNN
Oct. 2008	181.07	192.932	191.049	191.131	172.084
Nov. 2008	180.56	191.127	192.015	187.827	172.597
Dec. 2008	189.03	189.916	192.981	184.999	176.614
Jan. 2009	182.07	191.995	193.947	185.613	177.641
Feb. 2009	167.35	189.940	194.913	184.397	180.343
Mar. 2009	189.30	183.988	195.879	178.988	183.830
Apr. 2009	175.84	189.348	196.846	181.395	187.104
MAPE (%)		6.044	7.480	4.636	5.062

3.2.3 GRNN Model

For the GRNN model, Fig. 3.4 shows the MAPE values of the GRNN with various σ. Clearly, while σ exceeds over 3.33, the value of MAPE will subsequently increase. Therefore, the limit of σ is 3.33. In this book, the value of σ is set at 0.04. The forecasting results are shown in the fifth column of Table 3.3.

3.2.4 BPNN Model

For the BPNN model, Matlab 6.5 computing software is employed to implement the forecasting procedure. The number of nodes in the hidden layer is used as a validation parameter of the BPNN model. The most suitable number of hidden

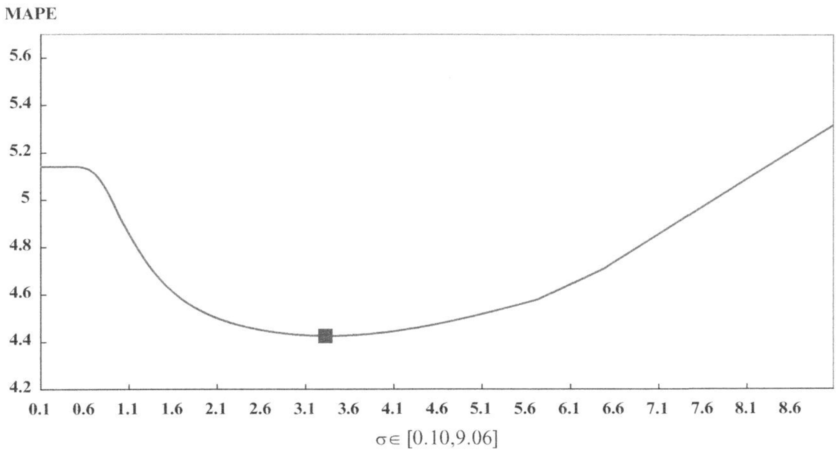

Fig. 3.4 MAPE with various σ values of GRNN model

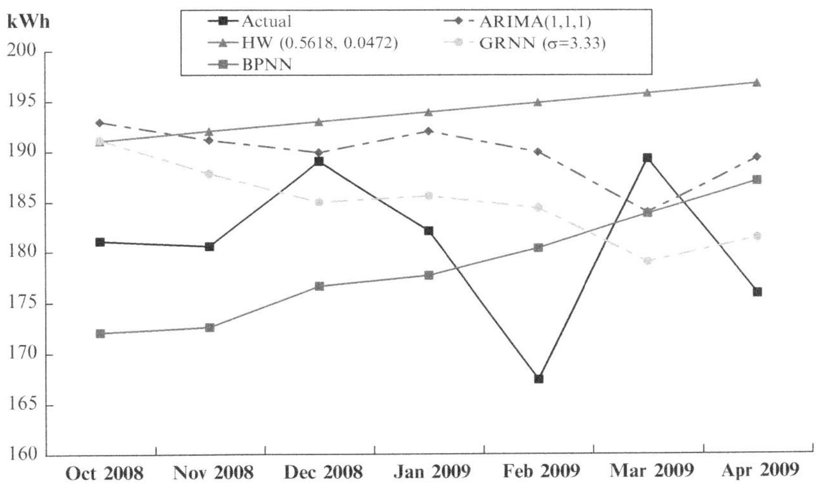

Fig. 3.5 Forecasting results of ARIMA, HW, GRNN, and BPNN models

nodes of the BPNN model is set as 3. The forecasting results are shown in the last column of Table 3.3.

Figure 3.5 is provided to illustrate the forecasting accuracy among different models. Obviously, these four models are not fitting well particularly for HW and BPNN models. Therefore, it is desired to look for more robust technology to overcome these shortcomings from the traditional models and ANN-based models. In the following sections, ARIMA(1,1,1) and GRNN(σ = 3.33) are selected to be compared with SVR-based models.

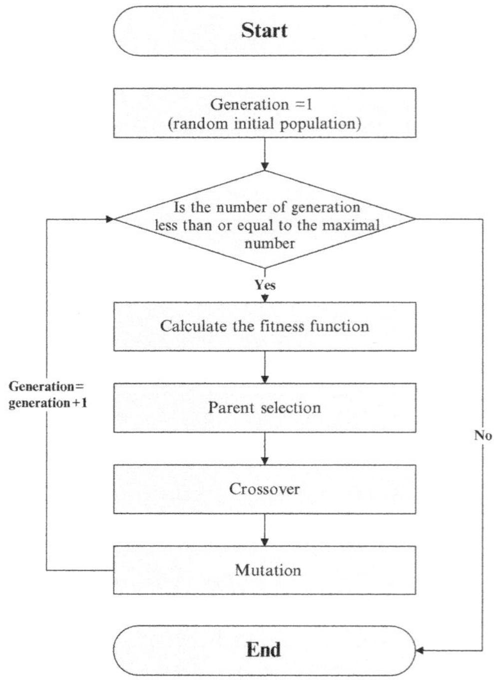

Fig. 3.6 The architecture of the genetic algorithm

3.3 Genetic Algorithm in SVR's Parameter Determination

3.3.1 Operation Procedure of GA

Proposed by Holland [7], genetic algorithm (GA) is an organized random search technique by imitating the biological evolution process. Such algorithm is based on the principle of the survival of the fittest, which attempts to retain genetic information from generation to generation. GA is also auto-adaptive stochastic search technique [7]; it generates new individuals with selection, crossover, and mutation operators. GA starts with a coding of the parameter set of all types of objective functions; thus, it has the ability to solve those traditional algorithms that are not easy to solve. The major advantages of GA are the capabilities for finding optimal or near optimal solutions with relatively modest computational requirements. Figure 3.6 depicts the operation of a GA, which is described below.

Step 1: Initialization. Generate randomly an initial population of chromosomes. The three free parameters, σ, C, and ε, are encoded in a binary format and represented by a chromosome.

Step 2: Evaluating fitness. Evaluate the fitness of each chromosome. In this book, a negative mean absolute percentage error (−MAPE) is used as the fitness function. The MAPE is as Eq. (3.13):

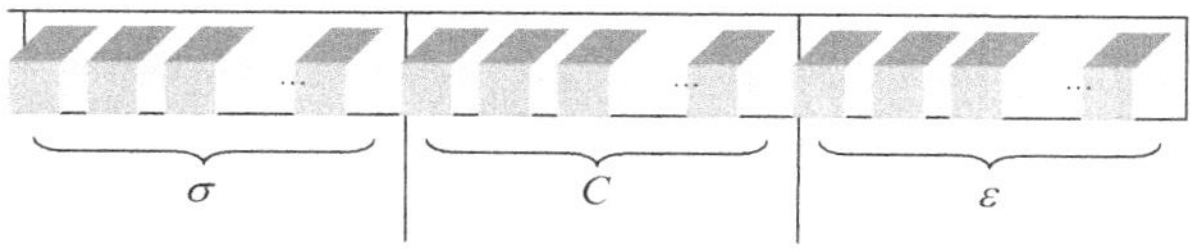

Fig. 3.7 The binary encoding of a chromosome

$$\text{MAPE} = \frac{1}{N}\sum_{i=1}^{N}\left|\frac{\text{a}_i - \text{f}_i}{\text{a}_i}\right| \times 100\,\%, \tag{3.13}$$

where a_i and f_i represent the actual electric load and forecast electric load, and N is the number of forecasting periods.

Step 3: Selection. Based on fitness functions, chromosomes with higher fitness values are more likely to yield offspring in the next generation. The roulette wheel selection principle [7] is applied to choose chromosomes for reproduction.

Step 4: Crossover and mutation. Create new offspring by crossover and mutation operations. Mutations are performed randomly by converting a "1" bit into a "0" bit or a "0" bit into a "1" bit. The single-point-crossover principle is employed. Segments of paired chromosomes between two determined breakpoints are swapped. The rates of crossover and mutation are probabilistically determined. In this book, the probabilities of crossover and mutation are set to 0.5 and 0.1, respectively.

Step 5: Next generation. Form a population for the next generation.

Step 6: Stop condition. If the number of generation is equal to a given scale, then the best chromosomes are presented as a solution; otherwise, go back to step 2.

GA is used to yield a smaller MAPE by searching for better combinations of three parameters in SVR. In this book, binary encoding is specified in GA. Three free parameters, σ, C and ε, are represented by a chromosome that consists of three genes in the form of binary numbers (Fig. 3.7). The size of the population is set to 200 herein. Each gene contains 40 bits. If each gene contains 40 bits, for example, then a chromosome contains 120 bits. More bits in a gene correspond to a finer partition of the searched space. Parent selection is a procedure in which two chromosomes from the parent population are chosen according to the fitness functions. Fitter chromosomes are more likely to generate offspring to the next generation. For simplicity, suppose a gene has four bits. A chromosome contains 12 bits (Fig. 3.8). Before crossover is performed, the values of the three parameters in #1 parent are 1.5, 1.25, and 0.34375. For #2 parent, the three values are 0.625, 8.75, and 0.15625. After crossover, for #1 offspring, the three values are 1.625, 3.75, and 0.40625. For #2 offspring, the three values are 0.5, 6.25, and 0.09375. Mutations are performed randomly by converting a "1" bit into a "0" bit or a "0" bit into a "1" bit. The rates of crossover and mutation are probabilistically determined. In this study, the probabilities of crossover and mutation are set to 0.5 and 0.1, respectively.

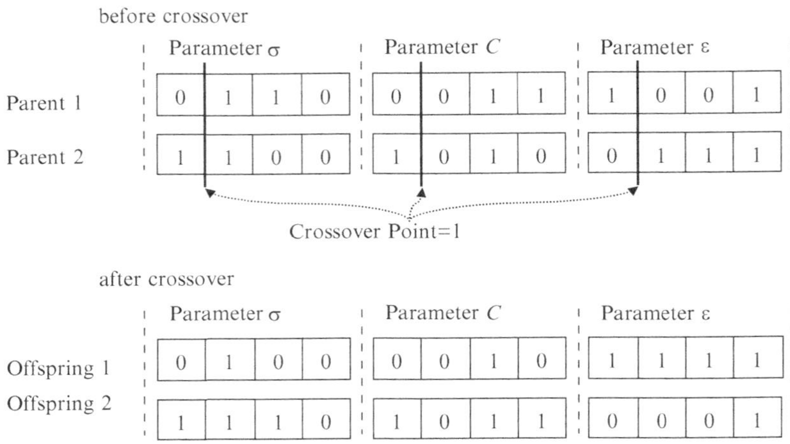

Fig. 3.8 A simplified example of parameter representation

3.3.2 GA for Three-Parameter Determination and Forecasting Results

This subsection will demonstrate the three-parameter determination of the proposed hybrid model (SVR with GA), namely, SVRGA model.

In the training stage, the rolling-based forecasting procedure (see Fig. 3.9) is conducted, which is dividing training data into two subsets, namely, fed-in subset (e.g., 25 load data) and fed-out subset (7 load data), respectively. Firstly, the primary 25 load data of fed-in subset are feeding into the proposed model (the structural risk minimization principle is employed to minimize the training error) and then obtain one-step-ahead forecasting load, namely, the 26th forecasting load. Secondly, the next 25 load data, including 24 of the fed-in subset data (from 2nd to 25th) pulsing the 26th data in the fed-out subset, are similarly again fed into the proposed model (the structural risk minimization principle is also employed to minimize the training error) and then obtain one-step-ahead forecasting load, namely, the 27th forecasting load. Repeat the rolling-based forecasting procedure till the 32nd forecasting load is obtained. Meanwhile, training error in this training stage is also obtained.

For the first training rolling, 25 months (from Dec. 2004 to Dec. 2006), electric load data are feeding into SVRGA to compute the forecasting electric load of Jan. 2007, then, for the second training rolling, the next 25 months (from Jan. 2005 to Jan. 2007), electric load data are similarly feeding into SVRGA to compute the forecasting electric load of Feb. 2007, and repeat previous procedures to compute the forecasting electric load of Mar. 2007,. . ., Jul. 2007, respectively. Finally, compute the MAPE of training stage between the forecasting electric load (from Jan. 2007 to Jul. 2007) and the original load (from Jan. 2007 to Jul. 2007, namely, fed-out subset).

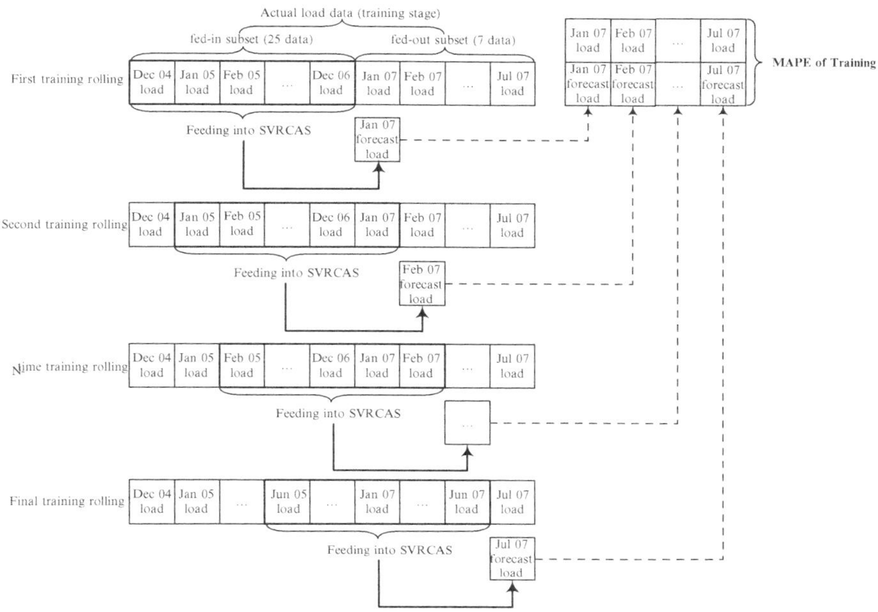

Fig. 3.9 The rolling-base forecasting procedure (training stage)

Table 3.4 Parameter determination of SVRGA model

	Parameters			
Nos. of fed-in data	σ	C	ε	MAPE of testing (%)
5	76.84	297.20	0.4298	4.354
10	4.46	143.54	1.6705	3.763
15	4.67	70.23	3.9921	3.719
20	233.56	2,911.70	11.2340	3.974
25	**686.16**	**5,048.40**	**19.3170**	**3.676**

While training error improvement occurs, the three kernel parameters, σ, C, and ε, of the SVRGA model adjusted by GA are employed to calculate the validation error. Then, the adjusted parameters with minimum validation error are selected as the most appropriate parameters. Finally, a one-step-ahead policy is employed to forecast electric load. Note that the testing data set is not used for modeling but for examining the accuracy of the forecasting model. The forecasting results and the suitable parameters for the SVRGA model are illustrated in Table 3.4, in which it is indicated that these two models all perform the best when 25 fed-in data are used.

Table 3.5 shows the actual values and the forecast values obtained using various forecasting models: ARIMA(1,1,1), GRNN(σ = 3.33), TF-ε-SVR-SA, and SVRGA. The MAPE values are calculated to compare fairly the proposed models with other alternative models. The proposed SVRGA model has smaller MAPE values than

Table 3.5 Forecasting results of ARIMA, GRNN, TF-ε-SVR-SA, and SVRGA models (unit, hundred million kWh)

Time point (month)	Actual	ARIMA(1,1,1)	GRNN($\sigma = 3.33$)	TF-ε-SVR-SA	SVRGA
Oct. 2008	181.07	192.932	191.131	184.504	178.326
Nov. 2008	180.56	191.127	187.827	190.361	178.355
Dec. 2008	189.03	189.916	184.999	202.980	178.355
Jan. 2009	182.07	191.995	185.613	195.753	178.356
Feb. 2009	167.35	189.940	184.397	167.580	178.357
Mar. 2009	189.30	183.988	178.988	185.936	178.358
Apr. 2009	175.84	189.348	181.395	180.165	181.033
MAPE (%)		6.044	4.636	3.799	3.676

Table 3.6 Wilcoxon signed-rank test

	Wilcoxon signed-rank test	
	$\alpha = 0.025$	$\alpha = 0.05$
Compared models	$W = 2$	$W = 3$
SVRGA vs. ARIMA(1,1,1)	0[a]	0[a]
SVRGA vs. GRNN($\sigma = 3.33$)	0[a]	0[a]
SVRGA vs. TF-ε-SVR-SA	2[a]	2[a]

[a]Denotes that SVRGA model significantly outperforms other alternative models

Table 3.7 Asymptotic test

	Asymptotic (S_1) test	
Compared models	$\alpha = 0.05$	$\alpha = 0.10$
SVRGA vs. ARIMA(1,1,1)	H_0: $e_1 = e_2$ $S_1 = -11.546$; p=0.000 (reject H_0)	H_0: $e_1 = e_2$ $S_1 = -11.546$; $p = 0.000$ (reject H_0)
SVRGA vs. GRNN($\sigma = 3.33$)	H_0: $e_1 = e_2$ $S_1 = -2.100$; $p = 0.0179$ (reject H_0)	H_0: $e_1 = e_2$ $S_1 = -2.100$; $p = 0.0179$ (reject H_0)
SVRGA vs. TF-ε-SVR-SA	H_0: $e_1 = e_2$ $S_1 = -2.344$; $p = 0.00954$ (reject H_0)	H_0: $e_1 = e_2$ $S_1 = -2.344$; $p = 0.00954$ (reject H_0)

ARIMA, GRNN, and TF-ε-SVR-SA models. Furthermore, to verify the significance of accuracy improvement of SVRGA model comparing with ARIMA(1,1,1), GRNN ($\sigma = 3.33$), and TF-ε-SVR-SA models, the statistical test, namely, Wilcoxon signed-rank test, is conducted at the 0.025 and 0.05 significance levels in one-tail tests.

The test results are shown in Tables 3.6 and 3.7, respectively. Clearly, the SVRGA model is significantly superior to ARIMA(1,1,1), GRNN($\sigma = 3.33$), and TF-ε-SVR-SA. Figure 3.10 is provided to illustrate the forecasting accuracy among different models.

The superior performance of the SVRGA model has several causes: first, for a SVR-based model, it has nonlinear mapping capabilities and can more easily

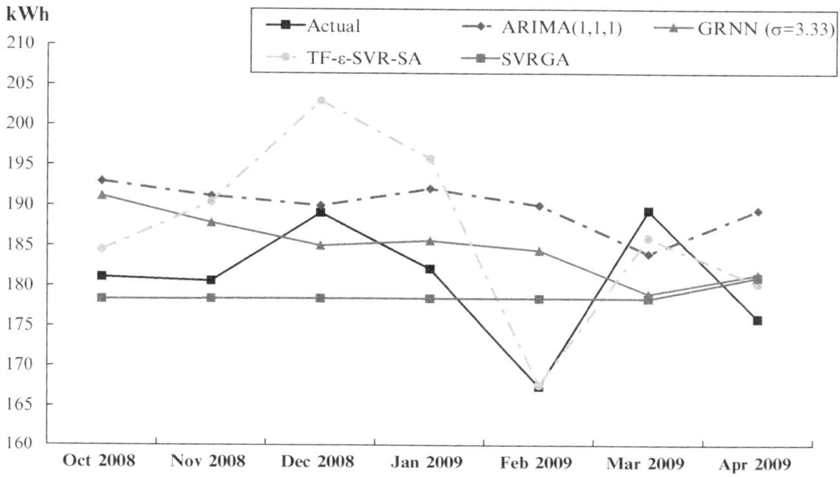

Fig. 3.10 Forecasting results of ARIMA, GRNN, TF-ε-SVR-SA, and SVRGA models

capture electric load data patterns than can the ARIMA and GRNN models. Second, the parameter selection in a SVR model significantly influences their forecasting performance. Improper determining of these three parameters will cause either over-fitting or under-fitting of a SVR model. In this section, the GA can determine suitable free parameters for forecasting electric load. Finally, the SVR-based model performs structural risk minimization rather than minimizing the training errors. Minimizing the upper bound on the generalization error improves the generalization performance compared to the ARIMA and GRNN models. However, it is clear that SVRGA is not fitting the actual electric loads very well even it has smaller MAPE values and passed the Wilcoxon test than other alternatives. Therefore, it still requires hybridizing other novel techniques to improve this shortcoming.

3.4 Simulated Annealing Algorithm in SVR's Parameter Determination

3.4.1 Operation Procedure of SA Algorithm

The simulated annealing (SA) algorithm is an optimization technique, analogous to the annealing process of material physics. Boltzmann [8] pointed out if the system is in thermal equilibrium at a temperature T, then the probability $P_T(s)$ of the system being in a given state s is given by the Boltzmann distribution, shown as Eq. (3.14):

$$P_T(s) = \frac{\exp(-E(s)/kT)}{\sum\limits_{w \in S} \exp(-E(w)/kT)}, \tag{3.14}$$

where $E(s)$ denotes the energy of state s (the state is defined as the SVR forecasting error in each iteration here), k represents the Boltzmann constant, and S is the set of all possible states. However, Eq. (3.14) does not contain information on how a fluid reaches thermal equilibrium at a given temperature. Metropolis et al. [9] develop an algorithm that simulates the process of Boltzmann. The Metropolis algorithm is summarized as follows. When the system is in original state s_{old} with energy $E(s_{\text{old}})$, a randomly selected atom is perturbed, resulting in a state s_{new} with energy $E(s_{\text{new}})$. This new state is either accepted or rejected depending on the Metropolis criterion: if $E(s_{\text{new}}) \leq E(s_{\text{old}})$, then the new state is automatically accepted; in contrast, if $E(s_{\text{new}}) > E(s_{\text{old}})$, then the probability of accepting the new state is given by the following probability function, Eq. (3.15):

$$P(\text{accept } s_{\text{new}}) = \exp\left(-\frac{E(s_{\text{old}}) - E(s_{\text{new}})}{kT}\right). \tag{3.15}$$

According to the studies of Boltzmann and Metropolis, Kirkpatrick et al. [10] claim that the Metropolis approach is conducted for each temperature on the annealing schedule until thermal equilibrium is reached. Additionally, a prerequisite for applying SA is that a given set of the multiple variables defines a unique system state, for which the objective function can be calculated. The procedure of SA algorithm is described as follows and the flowchart is shown as Fig. 3.11.

Step 1: Initialization. Set upper bounds of the three SVR parameters, σ, C, and ε. Then, generate and feed the initial values of the three parameters into the SVR model. The forecasting error is defined as the system state (E). Here, the initial state (E_0) is obtained.

Step 2: Provisional state. Make a random move to change the existing system state to a provisional state. Another set of three positive parameters is generated in this stage.

Step 3: Acceptance tests. The following equation is employed to determine the acceptance or rejection of the provisional state [9]:

$$\begin{cases} \text{Accept the provisional state,} & \text{if } E(s_{\text{new}}) > E(s_{\text{old}}), \\ \quad \text{and } p < P(\text{accept } s_{\text{new}}),\ 0 \leq p \leq 1 & \\ \text{Accept the provisional state,} & \text{if } E(s_{\text{new}}) \leq E(s_{\text{old}}) \\ \text{Reject the provisional state,} & \text{otherwise.} \end{cases} \tag{3.16}$$

In Eq. (3.16), the p is a random number for determining the acceptance of the provisional state. If the provisional state is accepted, then set the provisional state as the current state.

Step 4: Incumbent solutions. If the provisional state is not accepted, then return to step 2. Furthermore, if the current state is not superior to the system state, then repeat steps 2 and 3 until the current state is superior to the system state, and set the current state as the new system state. Previous studies [10, 11] indicate that the

Fig. 3.11 The architecture of the SA algorithm

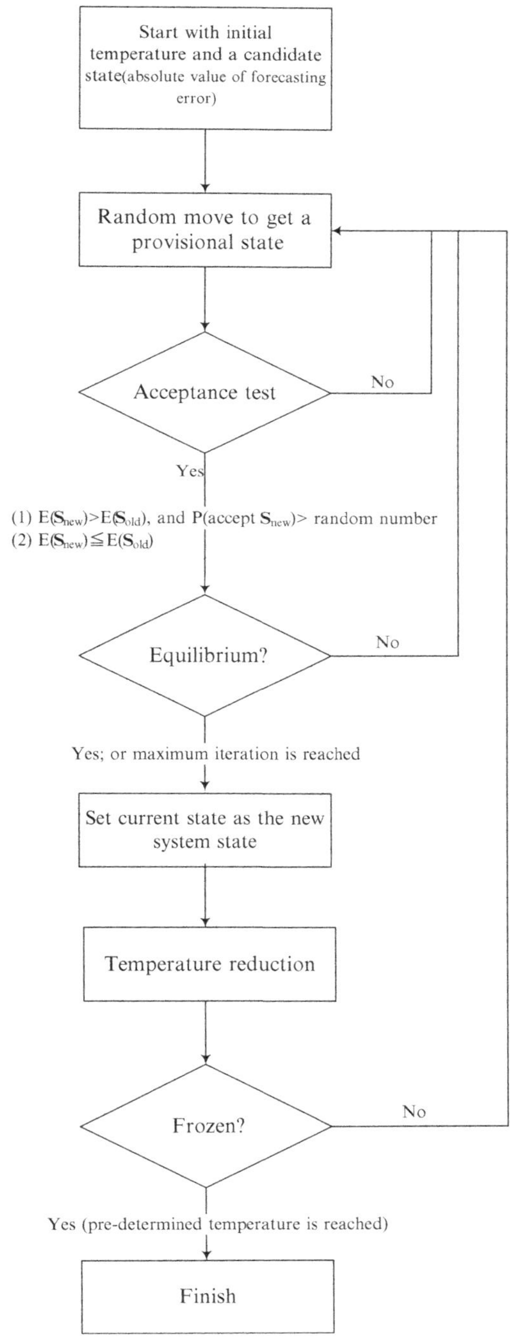

Table 3.8 Parameter determination of SVRSA model

Nos. of fed-in data	Parameters			MAPE of testing (%)
	σ	C	ε	
5	464.06	399.70	0.6891	4.289
10	3.72	176.14	0.6089	4.161
15	3.53	165.38	7.3935	3.941
20	3.02	1,336.70	9.8374	3.871
25	**94.998**	**9,435.20**	**12.6570**	**3.801**

maximum number of loops (N_{sa}) is 100d to avoid infinitely repeated loops, where d denotes the problem dimension. In this investigation, three parameters (σ, C, and ε) are used to determine the system states. Therefore, N_{sa} is set to be 300.

Step 5: Temperature reduction. After the new system state is obtained, reduce the temperature. The new temperature reduction is obtained by the Eq. (3.17),

$$\text{New temperature} = \text{Current temperature} \times \rho, \tag{3.17}$$

where $0 < \rho < 1$. The ρ is set to be 0.9 in this book [12]. If the predetermined temperature is reached, then stop the algorithm and the latest state is an approximate optimal solution. Otherwise, go to step 2.

Similarly, the value of the mean absolute percent error (MAPE), shown as Eq. (3.13), also serves as the criterion for identifying suitable parameters for use in the SVRSA model. The SA algorithm is used to seek a better combination of the three parameters in a SVR model, so that a smaller MAPE is obtained in each iteration.

3.4.2 *SA Algorithm for Three-Parameter Determination and Forecasting Results*

This subsection will demonstrate the three-parameter determination of the proposed hybrid model (SVR with SA), namely, SVRSA model.

Similarly, in the training stage, the rolling-based forecasting procedure is also employed to obtain the forecasting load and receive training error in the training stage. Then, if training error improvement occurs, the three kernel parameters, σ, C, and ε, of the SVRSA model adjusted by SA algorithm are employed to calculate the validation error. The adjusted parameters with minimum validation error are also selected as the most appropriate parameters. The forecasting results and the suitable parameters for the SVRSA model are illustrated in Table 3.8, in which it is also indicated that these two models all perform the best when 25 fed-in data are used.

Table 3.9 shows the actual values and the forecast values obtained using various forecasting models: ARIMA(1,1,1), GRNN($\sigma = 3.33$), TF-ε-SVR-SA, and SVRSA. The MAPE values are calculated to compare fairly the proposed models with other alternative models. The proposed SVRSA model only has smaller MAPE values than ARIMA and GRNN models, but TF-ε-SVR-SA model. Furthermore, to verify the

Table 3.9 Forecasting results of ARIMA, GRNN, TF-ε-SVR-SA, and SVRSA models (unit, hundred million kWh)

Time point (month)	Actual	ARIMA(1,1,1)	GRNN($\sigma = 3.33$)	TF-ε-SVR-SA	SVRSA
Oct. 2008	181.07	192.932	191.131	184.504	184.584
Nov. 2008	180.56	191.127	187.827	190.361	185.412
Dec. 2008	189.03	189.916	184.999	202.980	185.557
Jan. 2009	182.07	191.995	185.613	195.753	185.593
Feb. 2009	167.35	189.940	184.397	167.580	185.737
Mar. 2009	189.30	183.988	178.988	185.936	184.835
Apr. 2009	175.84	189.348	181.395	180.165	184.390
MAPE (%)		6.044	4.636	3.799	3.801

Table 3.10 Wilcoxon signed-rank test

	Wilcoxon signed-rank test	
	$\alpha = 0.025$	$\alpha = 0.05$
Compared models	$W = 2$	$W = 3$
SVRSA vs. ARIMA(1,1,1)	1[a]	1[a]
SVRSA vs. GRNN($\sigma = 3.33$)	2[a]	2[a]
SVRSA vs. TF-ε-SVR-SA	3	3[b]

[a]Denotes that SVRSA model significantly outperforms other alternative models
[b]Denotes that SVRSA model is significantly outperformed by other alternative models

Table 3.11 Asymptotic test

	Asymptotic (S_1) test	
Compared models	$\alpha = 0.05$	$\alpha = 0.10$
SVRSA vs. ARIMA(1,1,1)	H_0: $e_1 = e_2$ $S_1 = -9.790$; $p = 0.000$ (reject H_0)	H_0: $e_1 = e_2$ $S_1 = -9.790$; $p = 0.000$ (reject H_0)
SVRSA vs. GRNN($\sigma = 3.33$)	H_0: $e_1 = e_2$ $S_1 = -1.210$; $p = 0.1131$ (not reject H_0)	H_0: $e_1 = e_2$ $S_1 = -1.210$; $p = 0.1131$ (not reject H_0)
SVRSA vs. TF-ε-SVR-SA	H_0: $e_1 = e_2$ $S_1 = -0.969$; $p = 0.1663$ (not reject H_0)	H_0: $e_1 = e_2$ $S_1 = -0.969$; $p = 0.1663$ (not reject H_0)

significance of accuracy improvement of SVRSA model comparing with ARIMA (1,1,1), GRNN($\sigma = 3.33$), and TF-ε-SVR-SA models, the Wilcoxon signed-rank test and asymptotic test, as mentioned, are also conducted. The test results are shown in Tables 3.10 and 3.11, respectively. Clearly, the SVRSA model is significantly superior to ARIMA(1,1,1) model and minor significantly superior to GRNN ($\sigma = 3.33$) model (only receives significance with both levels in Wilcoxon test, but all fails with both levels in asymptotic test). And, based on Tables 3.9, 3.10, and 3.11, TF-ε-SVR- SA model has smaller MAPE value but not complete significantly (only receives significance with $\alpha = 0.05$ level in Wilcoxon test, but all fails with both levels in asymptotic test) outperforms SVRSA model. Figure 3.12 is provided to illustrate the forecasting accuracy among different models.

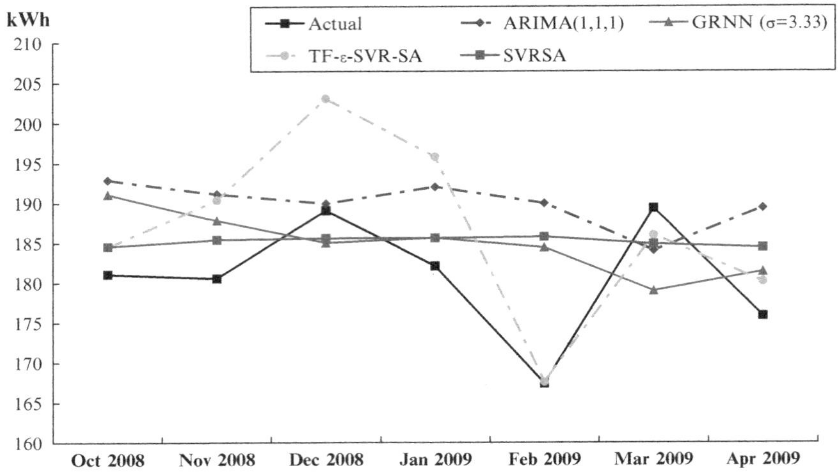

Fig. 3.12 Forecasting results of ARIMA, GRNN, TF-ε-SVR-SA, and SVRSA models

The superior performance of the SVRSA model not only caused from several similar causes of SVRGA model, such as SVR-based model with nonlinear mapping capabilities, minimizing the structural risks rather than the training errors, but also caused from the searching mechanism of SA algorithm itself. In this section, the SA algorithm can successfully escape from some critical local minimum (forecasting error) of the three-parameter combination for electric load forecasting. However, it is also clear that SVRSA model is not fitting the actual electric loads very well even if it has significant smaller MAPE values than other alternatives. Therefore, it also still requires hybridizing other novel techniques to improve this shortcoming.

3.5 Hybrid GA with SA in SVR's Parameter Determination

3.5.1 Shortcomings of GA and SA

GA is auto-adaptive stochastic search technique [7] that is based on the Darwinian survival-of-the-fittest philosophy and generates new individuals with selection, crossover, and mutation operators. GA starts with a coding of the parameter set of all types of objective functions; thus, GA has the ability to solve those traditional algorithms that are not easy to solve. GA is able to reserve a few best fitted members of the whole population for the next generation in the operation process; however, after some generations, GA may lead to a premature convergence to a local optimum in searching the suitable parameters of a SVR model.

SA is a stochastic based general search tool that mimics the annealing process of material physics [10]. When the system in the original state is greater than that of the new generated state, this new state is automatically accepted. In contrast, the new state is accepted by Metropolis criterion with a probability function. The performance of SA is dependent on the cooling schedule. Thus, SA has some institution to be able to escape from local minima and reach to the global minimum [13]. However, SA costs more computation time. To ensure the efficiency of SA, a proper temperature cooling rate (stop criterion) should be considered.

To overcome these drawbacks from GA and SA, it is necessary to find some effective approach and improvement to avoid misleading to the local optimum and to search optimum objective function efficiently. Genetic algorithm–simulated annealing (GA–SA) hybrid algorithm is a novel trial in dealing with the challenges mentioned above. The GA–SA can firstly employ the superiority of SA algorithm to escape from local minima and approximate to the global minimum, and secondly apply the mutation process of GA to improve searching ability in the range of values. So, the hybrid algorithm has been applied to the fields of system design [14], system and network optimization [15, 16], query to information retrieval system [17], continuous-time production planning [18, 19], and electrical power districting problem [20]. However, there is little application of the GA–SA to SVR's parameter determination. This investigation presented in this book is motivated by a desire to solve the problem of maintaining the premature convergence to a local optimum of GA and the efficiency of SA mentioned above in determining the three free parameters in a SVR electric load forecasting model, namely, SVRGASA.

3.5.2 Operation Procedure of GA–SA Algorithm

To overcome the drawbacks from GA and SA, this study proposes a hybrid GA–SA algorithm by applying the superiority of SA to escape from local minima and approximate to the global minimum, in addition, by using the mutation process of GA to improve searching ability in the range of values. On the other hand, to avoid computation executing time consuming, only the optimal individual of GA population will be delivered to the SA for further improving. The proposed GA–SA algorithm consists of the GA part and the SA part. GA evaluates the initial population and operates on the population using three basic genetic operators to produce new population (best individual); then, for each generation of GA, it will be delivered to SA for further processing. After finishing all the processes of SA, the modified individual will be sent back to GA for the next generation. These computing iterations will be never stopped till the termination condition of the algorithm is reached. The proposed procedure of GA–SA algorithm is illustrated as follow and the flowchart is shown as Fig. 3.13.

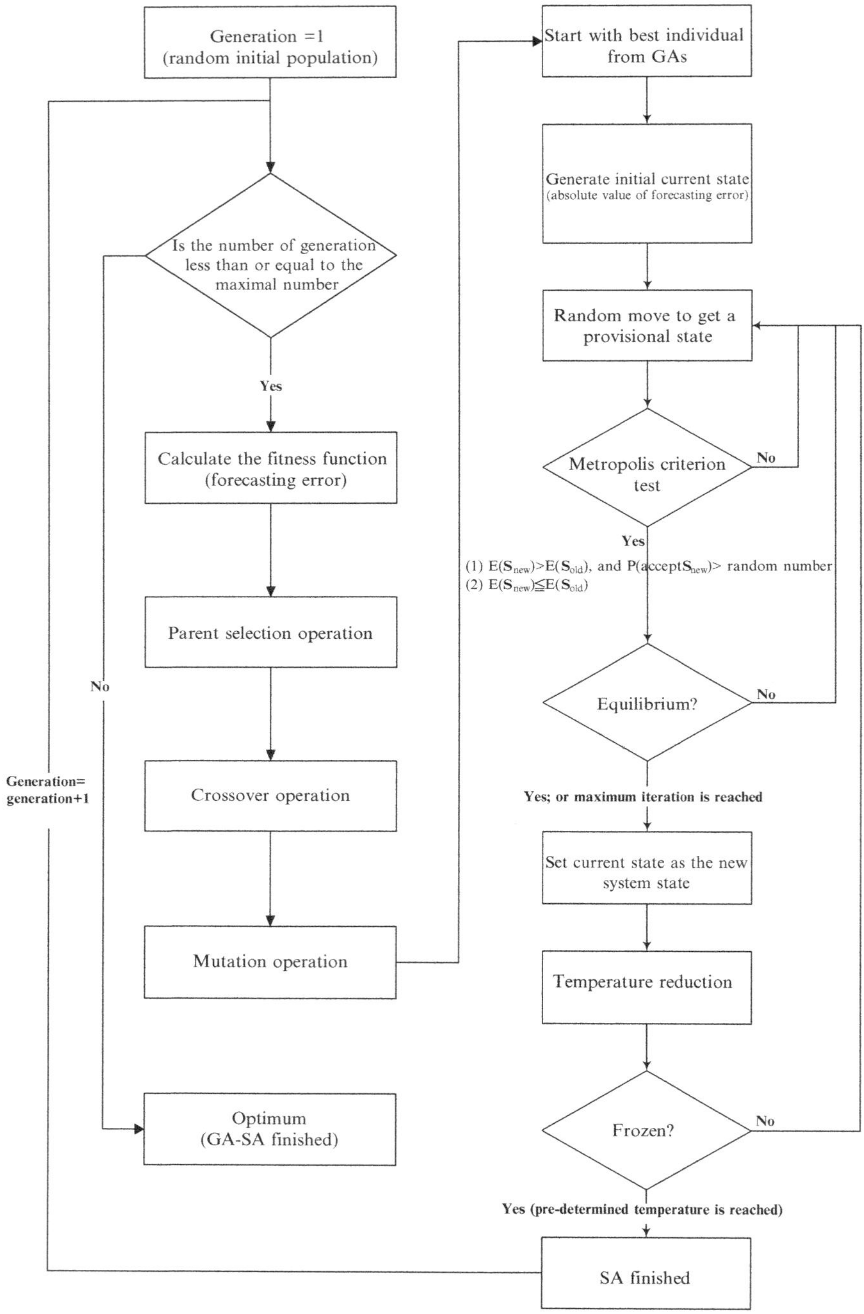

Fig. 3.13 The architecture of GA–SA algorithm

3.5.2.1 The Procedure of the GA Part

Step 1: Initialization. Construct randomly the initial population of chromosomes. The three parameters, C, σ, and ε, in a SVR model in the ith generation are encoded into a binary format and represented by a chromosome that is composed of "genes" of binary numbers (Fig. 3.2). Each chromosome has three genes, which represent three parameters. Each gene has 40 bits. For instance, if each gene contains 40 bits, a chromosome contains 120 bits. More bits in a gene correspond to finer partition of the search space.

Step 2: Evaluating fitness. Evaluate the fitness (forecasting errors) of each chromosome. In this book, a negative mean absolute percentage error (−MAPE) is used as the fitness function. The MAPE is as Eq. (3.13).

Step 3: Selection operation. Based on fitness functions, chromosomes with higher fitness values are more likely to yield offspring in the next generation. The roulette wheel selection principle [7] is also applied to choose chromosomes for reproduction.

Step 4: Crossover operation and mutation operation. Mutations are performed randomly by converting a "1" bit into a "0" bit or a "0" bit into a "1" bit. In crossover operation, chromosomes are paired randomly. The single-point-crossover principle is employed herein. Segments of paired chromosomes between two determined breakpoints are swapped. Finally, decode the crossover three parameters in a decimal format.

Step 5: Stop condition. If the number of generation is equal to a given scale, then the best chromosomes are presented as a solution; otherwise, go to the step 1 of the SA part.

In the proposed GA–SA algorithm process, GA will deliver its best individual to SA for further processing. After the optimal individual of GA being improved, SA sends it back to GA for the next generation. These computing iterations will be never stopped till the termination condition of the algorithm is reached.

3.5.2.2 The Procedure of the SA Part

Step 1: Generate initial current state. Receive values of the three parameters from GA. The value of forecasting error, MAPE, shown as Eq. (3.12), is defined as the system state (E). Here, the initial state (E_0) is obtained.

Step 2: Provisional state. Make a random move to change the existing system state to a provisional state. Another set of three positive parameters are generated in this stage.

Step 3: Metropolis criterion tests. Equation (3.16) is also employed to determine the acceptance or rejection of provisional state [9]. If the provisional state is accepted, then set the provisional state as the current state.

Step 4: Incumbent solutions. If the provisional state is not accepted, then return to step 2. Furthermore, if the current state is not superior to the system state, then repeat

Table 3.12 Parameter determination of SVRGASA model

	Parameters			
Nos. of fed-in data	σ	C	ε	MAPE of testing (%)
5	96.06	469.09	4.2588	5.049
10	22.45	99.97	0.9677	4.383
15	5.14	146.91	9.8969	3.951
20	788.75	6,587.20	9.2529	3.853
25	**92.09**	**2,449.50**	**13.639**	**3.530**

steps 2 and 3 until the current state is superior to the system state, and set the current state as new system state. The maximum number of loops (N_{sa}) is also set to 300.

Step 5: Temperature reduction. After the new system state is obtained, reduce the temperature. The new temperature reduction is obtained by the Eq. (3.17). If the predetermined temperature is reached, then stop the algorithm and the latest state is an approximate optimal solution. Otherwise, go to step 2.

3.5.3 GA–SA Algorithm for Three-Parameter Determination and Forecasting Results

This subsection will demonstrate the three-parameter determination of the proposed hybrid model (SVR with GA–SA), namely, SVRGASA model.

Similarly, in the training stage, the rolling-based forecasting procedure is also employed to obtain the forecasting load and receive training error in the training stage. Then, if training error improvement occurs, then the three kernel parameters, σ, C, and ε, of the SVRGASA model adjusted by GA–SA are employed to calculate the validation error. The adjusted parameters with minimum validation error are also selected as the most appropriate parameters. The forecasting results and the suitable parameters for the SVRGASA model are illustrated in Table 3.12, in which it is also indicated that these two models all perform the best when 25 fed-in data are used.

Table 3.13 shows the actual values and the forecast values obtained using various forecasting models: ARIMA(1,1,1), GRNN($\sigma = 3.33$), TF-ε-SVR-SA, SVRGA, SVRSA, and SVRGASA. The MAPE values are calculated to compare fairly the proposed models with other alternative models. The proposed SVRGASA model has smaller MAPE values than other alternative models. Furthermore, to verify the significance of accuracy improvement of SVRGASA model comparing with other alternative models, the Wilcoxon signed-rank test and asymptotic test, as mentioned, are also conducted. The test results are shown in Tables 3.14 and 3.15, respectively. Clearly, the SVRGASA model is significantly superior to other alternative models. Figure 3.14 is provided to illustrate the forecasting accuracy among different models.

Table 3.13 Forecasting results of ARIMA, GRNN, TF-ε-SVR-SA, SVRGA, SVRSA, and SVRGASA models (unit, hundred million kWh)

Time point (month)	Actual	ARIMA (1,1,1)	GRNN ($\sigma = 3.33$)	TF-ε-SVR-SA	SVRGA	SVRSA	SVRGASA
Oct. 2008	181.07	192.932	191.131	184.504	178.326	184.584	183.563
Nov. 2008	180.56	191.127	187.827	190.362	178.355	185.412	183.898
Dec. 2008	189.03	189.916	184.999	202.980	178.355	185.557	183.808
Jan. 2009	182.07	191.995	185.613	195.753	178.356	185.593	184.128
Feb. 2009	167.35	189.940	184.397	167.580	178.357	185.737	184.152
Mar. 2009	189.30	183.988	178.988	185.936	178.358	184.835	183.387
Apr. 2009	175.84	189.348	181.395	180.165	181.033	184.390	183.625
MAPE (%)		6.044	4.636	3.799	3.676	3.810	3.530

Table 3.14 Wilcoxon signed-rank test

	Wilcoxon signed-rank test	
	$\alpha = 0.025$	$\alpha = 0.05$
Compared models	$W = 2$	$W = 3$
SVRGASA vs. ARIMA(1,1,1)	0[a]	0[a]
SVRGASA vs. GRNN($\sigma = 3.33$)	2[a]	2[a]
SVRGASA vs. TF-ε-SVR-SA	1[a]	1[a]
SVRGASA vs. SVRGA	0[a]	0[a]
SVRGASA vs. SVRSA	0[a]	0[a]

[a]Denotes that SVRGASA model significantly outperforms other alternative models

Table 3.15 Asymptotic test

	Asymptotic (S_1) test	
Compared models	$\alpha = 0.05$	$\alpha = 0.10$
SVRGASA vs. ARIMA (1,1,1)	H_0: $e_1 = e_2$ $S_1 = -10.965$; $p = 0.000$ (reject H_0)	H_0: $e_1 = e_2$ $S_1 = -10.965$; $p = 0.000$ (reject H_0)
SVRGASA vs. GRNN ($\sigma = 3.33$)	H_0: $e_1 = e_2$ $S_1 = -1.879$; $p = 0.03016$ (reject H_0)	H_0: $e_1 = e_2$ $S_1 = -1.879$; $p = 0.03016$ (reject H_0)
SVRGASA vs. TF-ε-SVR-SA	H_0: $e_1 = e_2$ $S_1 = -2.432$; $p = 0.00751$ (reject H_0)	H_0: $e_1 = e_2$ $S_1 = -2.432$; $p = 0.00751$ (reject H_0)
SVRGASA vs. SVRGA	H_0: $e_1 = e_2$ $S_1 = 4.426$; $p = 0.000$ (reject H_0)	H_0: $e_1 = e_2$ $S_1 = 4.426$; $p = 0.000$ (reject H_0)
SVRGASA vs. SVRSA	H_0: $e_1 = e_2$ $S_1 = -17.370$; $p = 0.000$ (reject H_0)	H_0: $e_1 = e_2$ $S_1 = -17.370$; $p = 0.000$ (reject H_0)

In this section, the hybrid GA–SA algorithm helps to avoid trapping into local minimum than GA and SA algorithms do, thus, outperforming the SVRGA and SVRSA models. For example, in Tables 3.4, 3.8, and 3.12, the GA–SA algorithm is

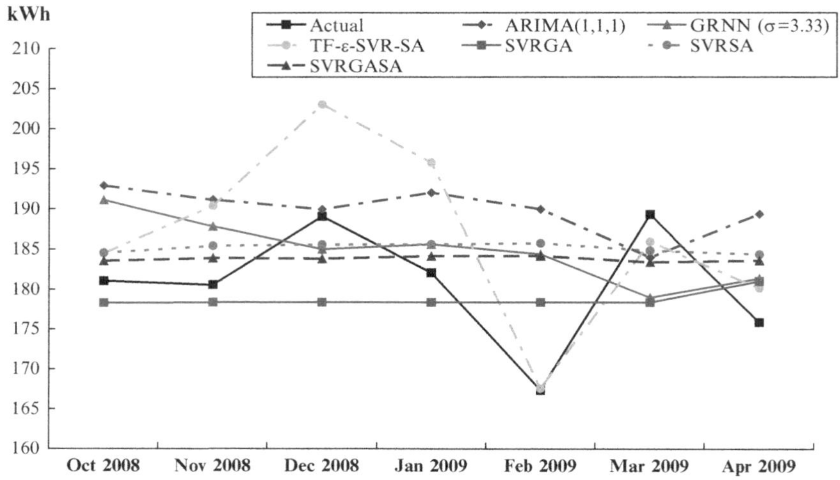

Fig. 3.14 Forecasting results of ARIMA, GRNN, TF-ε-SVR-SA, SVRGA, SVRSA, and SVRGASA models

excellently to shift the local solution of SVRGA and SVRSA models by 25 fed-in data rolling type, (σ, C, ε) = (686.16, 5048.4, 19.317 and 94.998, 9435.20, 12.6570, respectively) with local optimal forecasting errors, in terms of MAPE (3.676 % and 3.810 %, respectively), to be improved by GA–SA algorithm to another better solution, (σ, C, ε) = (92.807, 2449.50, 13.639) to be the appropriate local optimal forecasting error in terms of MAPE (3.530 %). Thus, it once again reveals that GA–SA algorithm is much appropriate than GA and SA algorithms in parameter adjustments to achieve forecasting accuracy improvement by integrated into the SVR model. However, it is also clear that SVRSA model is not fitting the actual electric loads very well even it has significant smaller MAPE values than other alternatives. Therefore, it also still requires hybridizing other novel techniques to improve this shortcoming.

3.6 Particle Swarm Optimization Algorithm in SVR's Parameter Determination

3.6.1 Operation Procedure of PSO Algorithm

In the previous sections, although both SVRGA and SVRSA are superior to other competitive forecasting models (ARIMA, HW, GRNN, and BPNN models), however, the drawbacks of GA and SA algorithms, as mentioned, are lacking knowledge memory or storage functions, while previous knowledge of the problem is destroyed once the population (GA) or the temperature changes (SA algorithm).

Thus, these drawbacks of GA and SA algorithms would lead to time-consuming and inefficient in searching the suitable parameters of a SVR model. Recently, inspired by the social behavior of organisms such as fish schooling and bird flocking, Kennedy and Eberhart [21] first introduced particle swarm optimization (PSO) algorithm, in which it is also initialized with a population of random solutions. Each individual, namely, particle, is assigned with a randomized velocity flown through hyperspace to look for the optimal position to land. Compared with GA and SA algorithm, PSO algorithm has memory to store the knowledge of good solutions by all particles; in addition, particles in the swarm share information with each other. Therefore, due to the simple concept, easy implementation, and quick convergence, nowadays PSO algorithm has gained much attention and wide applications in solving continuous nonlinear optimization problems [22]. However, the performance of PSO algorithm greatly depends on its parameters, and similar to GA and SA algorithm, it often suffers from being trapped in local optimum [23, 24].

In a PSO system, it starts with the random initialization of a population (swarm) in the search space, where multiple solution-candidates coexisted. Each solution, namely, particle, flies in the space looking for the optimal *position* to land. Eventually, the global best position of the system could be found out by adjusting the direction of each particle towards its own best location and towards the best particle of the swarm at each generation. The direction of each particle is adjusted by dynamically altering the *velocity* of each particle, according to its own flying experience as well as the experience of neighboring particles. During the searching process, tracking and memorizing the best position encountered could cumulate each particle's experience. Thus, the PSO system essentially has the capability of memory; each particle remembers the best position it reaches during the past, and then the PSO system combines local search method (via self experience) with global search methods (via neighboring experience).

The position, the velocity, and own best position of the ith particle pair, due to the three parameters in a SVR model, in the n-dimensional space can be represented as Eqs. (3.18)–(3.20), respectively,

$$X_{(k)i} = \left[x_{(k)i,1}, x_{(k)i,2}, \ldots, x_{(k)i,n}\right], \tag{3.18}$$

$$V_{(k)i} = \left[v_{(k)i,1}, v_{(k)i,2}, \ldots, v_{(k)i,n}\right], \tag{3.19}$$

$$P_{(k)i} = \left[p_{(k)i,1}, p_{(k)i,2}, \ldots, p_{(k)i,n}\right], \tag{3.20}$$

where $k = \sigma, C, \varepsilon$, and $i = 1, 2, \ldots, N$.

The global best position among all particles in the swarm $\mathbf{X}_{(k)i} = \left[X_{(k)1}, X_{(k)2}, \ldots, X_{(k)N}\right]$ is shown as Eq. (3.21):

$$P_{(k)g} = \left[p_{(k)g,1}, p_{(k)g,2}, \ldots, p_{(k)g,d}\right], \tag{3.21}$$

where $k = \sigma, C, \varepsilon$, and $g = 1, 2, \ldots, N$.

Then, the new velocity of each particle is computed by Eq. (3.22):

$$V_{(k)i}(t+1) = lV_{(k)i}(t) + q_1\text{rand}(\cdot)(P_{(k)i} - X_{(k)i}(t)) + q_2\text{Rand}(\cdot)(P_{(k)g} - X_{(k)i}(t)), \tag{3.22}$$

where $k = \sigma, C, \varepsilon$, and $i = 1, 2, \ldots, N$, l is called the inertia weight that controls the impact of the previous velocity of the particle on its current one, q_1 and q_2 are two positive constants called acceleration coefficients, and rand(·) and Rand(·) are two independently uniformly distributed random variables with range [0,1].

After the velocity has been updated, the new position of the particle for each parameter in the next generation is determined as Eq. (3.23):

$$X_{(k)i}(t+1) = X_{(k)i}(t) + V_{(k)i}(t+1), \tag{3.23}$$

where $k = \sigma, C, \varepsilon$, and $i = 1, 2, \ldots, N$.

Notice that the value of each component in $V_{(k)i}$ can be limited to the range $[-v_{\max}, v_{\max}]$ to control excessive roaming of particles outside the search space. This process is repeated until the defined stopping threshold is reached. The procedure of PSO algorithm is illustrated as follows and the flowchart is shown as Fig. 3.15. Interested readers could refer to [21] for more detail:

Step 1: Initialization. Initialize a defined population of particle pairs $(\sigma_i, C_i, \varepsilon_i)$ with random positions $(X_{\sigma i}, X_{Ci}, X_{\varepsilon i})$ and velocities $(V_{\sigma i}, V_{Ci}, V_{\varepsilon i})$, where each particle contains n variables.

Step 2: Objective value computing. Compute the objective values (forecasting errors) of all particle pairs. Let own best position $(P_{\sigma i}, P_{Ci}, P_{\varepsilon i})$ of each particle pair and its objective value $f_{\text{best}i}$ equal to its initial position and objective value. Let global best position $(P_{\sigma g}, P_{Cg}, P_{\varepsilon g})$ and its objective value $f_{\text{globalbest}i}$ equal to the best initial particle pair's position and its objective value.

Step 3: Evaluation of the objective values. According to Eqs. (3.22) and (3.23), update the velocity and position for each particle pair. And evaluate the objective values for all particle pairs.

Step 4: Comparison and update. For each particle pair, compare its current objective value with $f_{\text{best}i}$. If the current objective value is better (i.e., with smaller forecasting accuracy index value), then update best position $(P_{\sigma i}, P_{Ci}, P_{\varepsilon i})$ and its objective value by the current position and objective value.

Step 5: The best particle pair determination. Determine the best particle pair of whole population based on the best objective value. If the objective value is smaller than $f_{\text{globalbest}i}$, then update $(P_{\sigma g}, P_{Cg}, P_{\varepsilon g})$ and its objective value with the current best particle pair's position and objective.

Step 6: Stop criterion. If a stopping threshold (forecasting accuracy) is reached, then $(P_{\sigma g}, P_{Cg}, P_{\varepsilon g})$ and its $f_{\text{globalbest}i}$ would be determined; otherwise, go back to step 3.

Fig. 3.15 The architecture of PSO algorithm

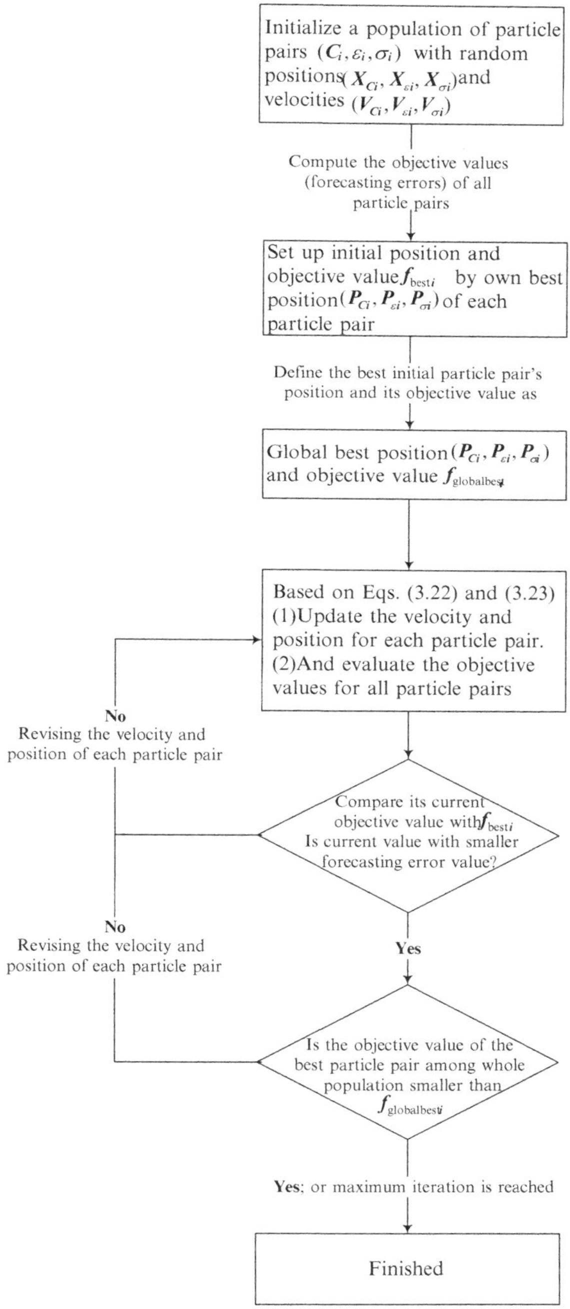

3.6.2 PSO Algorithm for Three-Parameter Determination and Forecasting Results

This subsection will demonstrate the three-parameter determination of the proposed hybrid model (SVR with PSO), namely, SVRPSO model.

The parameters of the PSO algorithm in the proposed model are experimentally set as shown in Table 3.16. The population size is set as 20; the total number of function evaluation is fixed as 10,000; q_1 and q_2 for each particle pair (σ, C, ε) are set to 0.05, 100, 0.5, respectively. $v_{\max}$ for σ particle is clamped to be 10 % of its search space (where $\sigma \in [0, 500]$). $v_{\max}$ for C particle is clamped to be 12.5 % of its search space ($C \in [0, 20000]$). $v_{\max}$ for ε particle are both clamped to be 15 % of its search space ($\varepsilon \in [0, 100]$).The standard PSO [21] uses a linearly varying inertia weight over the generations, varying from 1.2 at the beginning of the search to 0.2 at the end.

Similarly, in the training stage, the rolling-based forecasting procedure is also employed to obtain the forecasting load and receive training error in the training stage. Then, if training error improvement occurs, the three kernel parameter s, σ, C, and ε, of the SVRPSO model adjusted by PSO algorithm are employed to calculate the validation error. The adjusted parameters with minimum validation error are also selected as the most appropriate parameters. The forecasting results and the suitable parameters for the SVRPSO model are illustrated in Table 3.17, in which it is also indicated that these two models all perform the best when 25 fed-in data are used.

Table 3.18 shows the actual values and the forecast values obtained using various forecasting models: ARIMA(1,1,1), GRNN($\sigma = 3.33$), TF-ε-SVR-SA, and SVRPSO. The MAPE values are calculated to compare fairly the proposed models with other alternative models. The proposed SVRPSO model has smaller MAPE values than other alternative models. Furthermore, to verify the significance of accuracy improvement of SVRPSO model comparing with other alternative models, the Wilcoxon signed-rank test and asymptotic test are also conducted. The test results are shown in Tables 3.19 and 3.20, respectively. Clearly, the SVRPSO model is significantly superior to other alternative models, except versus GRNN model (receives significance with both levels in Wilcoxon test, but all fails with $\alpha = 0.05$ level in asymptotic test). Figure 3.16 is provided to illustrate the forecasting accuracy among different models.

In this section, the PSO algorithm is employed to overcome the shortcomings of GA and SA algorithms, that is, with memory to store the knowledge of good solutions and sharing information with each other. PSO algorithm is capable of searching experience memorizing to avoid inefficacious searching paths and to quickly converge. Therefore, it is expected to receive better forecasting performance than SVRGA and SVRSA models. However, it is also clear that SVRPSO model is not fitting the actual electric loads very well even if it has significant smaller MAPE values than other alternatives. Therefore, it also still requires hybridizing other novel techniques to improve this shortcoming.

Table 3.16 Parameters setting of PSO algorithm

Population size	Maximal iteration	Velocity limit						Inertia weight l		Acceleration coefficients					
		$-v_{max}$			v_{max}					q_1			q_2		
		σ	C	ε	σ	C	ε	l_{min}	l_{max}	σ	C	ε	σ	C	ε
20	10,000	−0.5	−2,500	−15	0.5	2,500	15	0.2	1.2	0.05	100	0.5	0.05	100	0.5

Table 3.17 Parameter determination of SVRPSO model

Nos. of fed-in data	Parameters σ	C	ε	MAPE of testing (%)
5	70.34	289.53	2.4341	4.558
10	23.82	81.12	1.2436	4.346
15	111.04	3,158.10	2.8713	4.484
20	93.32	5,683.70	11.4980	4.078
25	**158.44**	**7,014.50**	**2.2836**	**3.638**

Table 3.18 Forecasting results of ARIMA, GRNN, TF-ε-SVR-SA, SVRGA, SVRSA, and SVRPSO models (unit, hundred million kWh)

Time point (month)	Actual	ARIMA (1,1,1)	GRNN ($\sigma = 3.33$)	TF-ε-SVR-SA	SVRGA	SVRSA	SVRPSO
Oct. 2008	181.07	192.932	191.131	184.504	178.326	184.584	184.042
Nov. 2008	180.56	191.127	187.827	190.361	178.355	185.412	183.577
Dec. 2008	189.03	189.916	184.999	202.980	178.355	185.557	183.471
Jan. 2009	182.07	191.995	185.613	195.753	178.356	185.593	184.210
Feb. 2009	167.35	189.940	184.397	167.580	178.357	185.737	184.338
Mar. 2009	189.30	183.988	178.988	185.936	178.358	184.835	183.725
Apr. 2009	175.84	189.348	181.395	180.165	181.033	184.390	184.529
MAPE (%)		6.044	4.636	3.799	3.676	3.810	3.638

Table 3.19 Wilcoxon signed-rank test

Compared models	Wilcoxon signed-rank test $\alpha = 0.025$ $W = 2$	$\alpha = 0.05$ $W = 3$
SVRPSO vs. ARIMA(1,1,1)	0[a]	0[a]
SVRPSO vs. GRNN($\sigma = 3.33$)	2[a]	2[a]
SVRPSO vs. TF-ε-SVR-SA	2[a]	2[a]
SVRPSO vs. SVRGA	0[a]	0[a]
SVRPSO vs. SVRSA	1[a]	1[a]

[a]Denotes that SVRPSO model significantly outperforms other alternative models

3.7 Continuous Ant Colony Optimization Algorithm in SVR's Parameter Determination

3.7.1 *Basic Concept of ACO Algorithm*

Ant colony optimization (ACO) algorithm was firstly proposed by Dorigo [25] and Dorigo et al. [26]. The process by which ants could establish the shorter path between ant nest and food is illustrated in Fig. 3.17. Initially, ants leave their nest in random directions to search for food. As roaming around, ants deposit some

Table 3.20 Asymptotic test

Compared models	Asymptotic (S_1) test $\alpha = 0.05$	$\alpha = 0.10$
SVRPSO vs. ARIMA(1,1,1)	H_0: $e_1 = e_2$ $S_1 = -9.677$; $p = 0.000$ (reject H_0)	H_0: $e_1 = e_2$ $S_1 = -9.677$; $p = 0.000$ (reject H_0)
SVRPSO vs. GRNN($\sigma = 3.33$)	H_0: $e_1 = e_2$ $S_1 = -1.567$; $p = 0.0586$ (not reject H_0)	H_0: $e_1 = e_2$ $S_1 = -1.567$; $p = 0.0586$ (reject H_0)
SVRPSO vs. TF-ε-SVR-SA	H_0: $e_1 = e_2$ $S_1 = -1.852$; $p = 0.0320$ (reject H_0)	H_0: $e_1 = e_2$ $S_1 = -1.852$; $p = 0.0320$ (reject H_0)
SVRPSO vs. SVRGA	H_0: $e_1 = e_2$ $S_1 = 5.863$; $p = 0.000$ (reject H_0)	H_0: $e_1 = e_2$ $S_1 = 5.863$; $p = 0.000$ (reject H_0)
SVRPSO vs. SVRSA	H_0: $e_1 = e_2$ $S_1 = -5.992$; $p = 0.000$ (reject H_0)	H_0: $e_1 = e_2$ $S_1 = -5.992$; $p = 0.000$ (reject H_0)

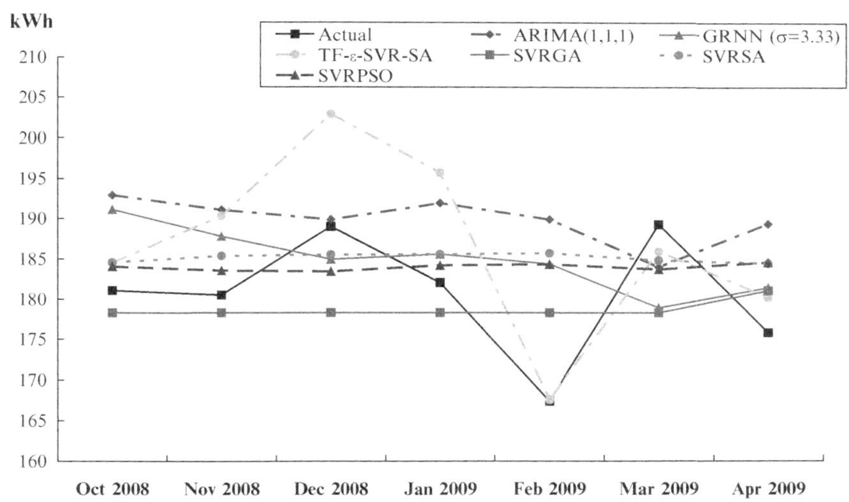

Fig. 3.16 Forecasting results of ARIMA, GRNN, TF-ε-SVR-SA, SVRGA, SVRSA, and SVRPSO models

amount of pheromone trails, which could be detectable by other ants. For example, assuming ant 1 finds a food source, it will pick up some food and go back to the nest by following its previous pheromone trail, laying *additional* pheromone on the same path while other ants (ant 2, ant 3, etc.) are still roaming randomly. When the second ant group leaves the nest to look for food, those ants could detect much

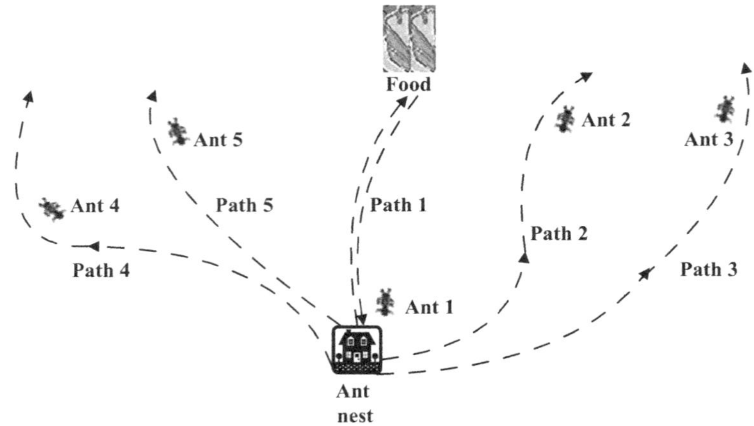

Fig. 3.17 Description of establishing the shorter path between ant nest and food

pheromone (twice) on path 1 than on others (path 2, path 3, etc.). Since the probability for a path to be followed is determined by its pheromone amount, more ants will follow path 1 in this second round of looking for food. In this way, the ants can establish the shortest path from their colony to the food sources. Obviously, even if an isolated ant roams randomly, it can, communicated by pheromones, follow the collective behavior of ant colonies.

Due to their power to learn and search capabilities, ACO algorithm has been successfully used to dealing with different combinatorial optimization problems including job-shop scheduling [27], traveling salesman problem [28], space planning [29], quadratic assignment problems [30], and data mining [31]. ACO imitates the behaviors of real ant colonies as they forage for food, wherein each ant lays down the pheromone on the path to the food sources or back to the nest. The paths with more pheromone are more likely to be selected by other ants. Over time, a colony of ants will select the shortest path to the food source and back to the nest. Therefore, a pheromone trail is the most important process for individual ant to smell and select its route. Meanwhile, ACO algorithm is originally proposed for discrete optimization, and their application to continuous optimization problems requires some specified transformation techniques. In the literature, only a few approaches for continuous optimization have been proposed, such as continuous ACO [32–34], API algorithm [35], and continuous interacting ACO [36]. However, these algorithms added some operational mechanisms that are mostly beyond the regular essences of ACO. Recently, Socha and Dorigo [37] proposed an extension of ACO to continuous domain by applying the continuous probability density function to decide pheromone probabilistic choice, in which, however, other external parameters should be determined in advance; thereby, it would mislead to continuous technological issue instead of appropriate SVR's parameter determination.

3.7.2 Continuing Transformation

Hence, the concepts of transforming a continuous search space to a discrete one by discretization of the continuous decision variables [38] are feasible to be employed, which is so-called the continuous ant colony optimization (CACO) algorithm. In this book, the CACO algorithm for the traveling salesman problem is modified to determine three parameters of a SVR model in the discrete search space. The probability, $P_k(i,j)$, that an ant k moves from city i to city j is expressed as Eq. (3.24):

$$P_k(i,j) = \begin{cases} \arg\max_{S\in M_k}\left\{[\tau(i,S)]^{\alpha}[\eta(i,S)]^{\beta}\right\}, & \text{if } q \le q_0 \\ \text{Eq.}(3.25), & \text{otherwise} \end{cases}, \quad (3.24)$$

$$P_k(i,j) = \begin{cases} [\tau(i,j)]^{\alpha}[\eta(i,j)]^{\beta} \Big/ \sum_{S\in M_k} [\tau(i,S)]^{\alpha}[\eta(i,S)]^{\beta}, & j \notin M_k \\ 0, & \text{otherwise} \end{cases}, \quad (3.25)$$

where $\tau(i,j)$ is the pheromone level between city i and city j, and $\eta(i,j)$ is the inverse of the distance between cities i and j. In this study, the forecasting error represents the distance between cities. The α and β are parameters determining the relative importance of pheromone level, and M_k is a set of cities in the next column of the city matrix for ant k. q is a random uniform variable [0,1], and the value q_0 is a constant between 0 and 1, that is, $q_0 \in [0,1]$. The values of α, β, and q_0 are set to be 8, 5, and 0.2, respectively.

Once ants have completed their tours, the most pheromone deposited by ants on the visited paths is considered as the information regarding the best paths from the nest to the food sources. Therefore, the pheromone dynamic updating plays the main role in real ant colonies searching behaviors. The local and global updating rules of pheromone are expressed as Eqs. (3.26) and (3.27), respectively,

$$\tau(i,j) = (1-\rho)\tau(i,j) + \rho\tau_0, \quad (3.26)$$

$$\tau(i,j) = (1-\delta)\tau(i,j) + \delta\Delta\tau(i,j), \quad (3.27)$$

where ρ is the local evaporation rate of pheromone, $0 < \rho < 1$; τ_0 is the initial amount of pheromone deposited on each of the paths. In this book, the value of ρ is set to be 0.01. In addition, the initial amount of pheromone, τ_0, generated by Dorigo and Gambardella's [28] proposed approach, is expressed as Eq. (3.28):

$$\tau_0 = \frac{1}{nL_{\mathrm{nn}}} \quad (3.28)$$

where n is the number of cities and L_{nn} is the tour length produced by the nearest neighbor heuristic.

Global trail updating is accomplished according to Eq. (3.27). The δ is the global pheromone decay parameter, $0 < \delta < 1$, and set to be 0.2 for this study. The $\Delta\tau(i, j)$, expressed as Eq. (3.29), is used to increase the pheromone on the path of the solution:

$$\Delta\tau(i,j) = \begin{cases} 1/L, & \text{if } (i,j) \in \text{global best route} \\ 0, & \text{otherwise} \end{cases}, \tag{3.29}$$

where L is the length of the shortest route.

3.7.3 Operation Procedure of CACO Algorithm

More detail of the CACO algorithm on this book is as follows and the flowchart is shown as Fig. 3.18.

Step 1: Initialization. Set upper bounds of three SVR positive parameters, σ, C, and ε. In this study, to discretize those continuous parameters, each digit of the parameters is represented by ten cities. Thus, each digit contains 10 possible values from 0 to 9. Assume the limits of parameters σ, C, and ε are 500, 10,000, and 100, correspondingly. The numbers of digits that represent each parameter (σ, C, and ε) are all set as six. Hence, three ant colonies are defined as σ-ant colony, C-ant colony, and ε-ant colony for three-parameter values searching. The numbers of cities for each ant colony are 40, and the total number of cities is 120.

Step 2: Assigning tasks to each ant colony. From step one, pathway-structure list of each ant colony would be generated. Figure 3.19 shows the parameters represented by the CACO algorithms and pathway-structure list in this study. Each ant will randomly select a pathway from the pathway list in its associate colony and remember the values of the represented parameters (σ, C, and ε). At the end of the pathway, pass the three-parameter values into the SVR model (i.e., objective function) and calculate the forecasting error. The shortest travel pathway in each searching loop would be determined based on the smaller forecasting error. In this book, the MAPE is used as the forecasting error index as given by Eq. (3.13).

Step 3: Determine the numbers of ants and calculate the distance between cities. The numbers of ants are set to be 10 in each ant colony searching, that is, totally 30 ants for each iteration searching. The maximum number of iterations is set to 20,000 to avoid infinite iterations.

Step 4: Stop criterion I. While the maximum number of iterations is reached, then stop the algorithm, and the shortest travel path of the ants colony is an approximate optimal solution. Otherwise, continue to step 4.

Step 5: Calculate the visiting probability. If the maximum number of iterations is not reached, then calculate the probability that an ant k in city i moves to city j in accordance with Eq. (3.24). Repeat steps 1–3.

Fig. 3.18 The architecture of CACO algorithm

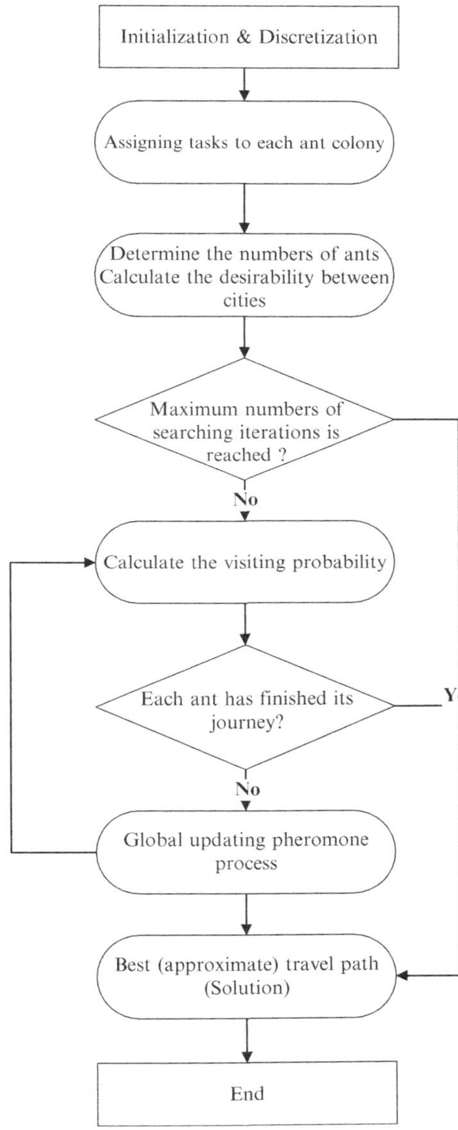

Step 6: Stop criterion II. If each ant has finished its pathway-structure list from the nest to the food source passing through all cities, then the shortest path is an approximate optimal solution. Otherwise, conduct the pheromone updating process represented as Eqs. (3.26) and (3.27) to renew the reinforcement of pheromone. Then, go back to step 4.

Notice that, in any iteration, while the shorter path is attained, the appropriate solution is determined, and for those three parameters, new search space is then

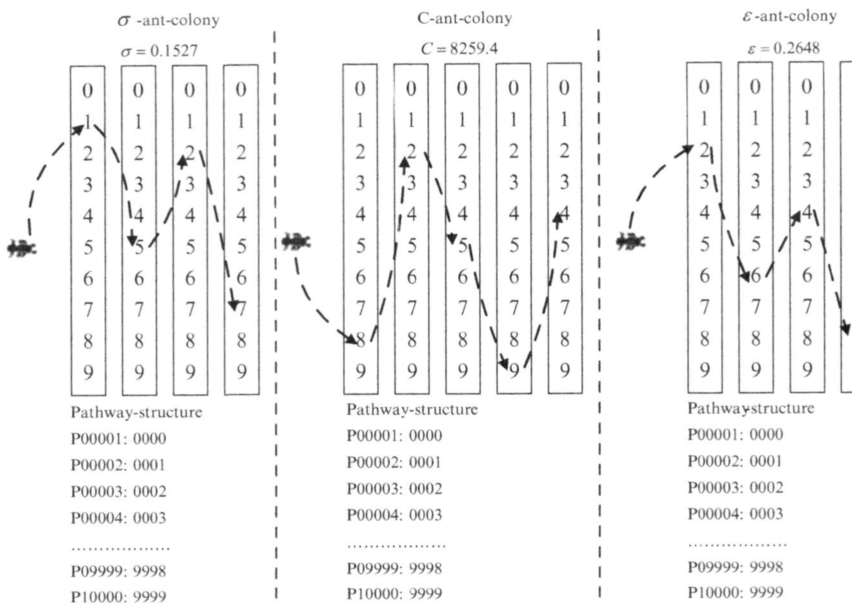

Fig. 3.19 SVR parameter representation by the CACO algorithm

re-discretized. The CACO algorithm is used to seek a better combination of the three parameters in the SVR so that a smaller MAPE is obtained during forecasting iteration.

3.7.4 CACO Algorithm for Three-Parameter Determination and Forecasting Results

This subsection will demonstrate the three-parameter determination of the proposed hybrid model (SVR with CACO), namely, SVRCACO model.

Similarly, in the training stage, the rolling-based forecasting procedure is also employed to obtain the forecasting load and receive training error in the training stage. Then, if training error improvement occurs, the three kernel parameters, σ, C, and ε, of the SVRCACO model adjusted by CACO algorithm are employed to calculate the validation error. The adjusted parameters with minimum validation error are also selected as the most appropriate parameters. The forecasting results and the suitable parameters for the SVRCACO model are illustrated in Table 3.21, in which it is also indicated that these two models all perform the best when 25 fed-in data are used.

Table 3.22 shows the actual values and the forecast values obtained using various forecasting models: ARIMA(1,1,1), GRNN(σ = 3.33), TF-ε-SVR-SA, SVRPSO, and SVRCACO. The MAPE values are calculated to compare fairly the proposed

Table 3.21 Parameter determination of SVRCACO model

Nos. of fed-in data	Parameters σ	C	ε	MAPE of testing (%)
5	1.49	322.92	6.7778	5.623
10	159.76	198.03	4.5219	5.076
15	12.81	114.24	0.0035	4.510
20	22.99	7,233.00	13.7640	4.003
25	**243.55**	**6,868.10**	**11.2480**	**3.371**

Table 3.22 Forecasting results of ARIMA, GRNN, TF-ε-SVR-SA, SVRPSO, and SVRCACO models (unit, hundred million kWh)

Time point (month)	Actual	ARIMA (1,1,1)	GRNN ($\sigma = 3.33$)	TF-ε-SVR-SA	SVRPSO	SVRCACO
Oct. 2008	181.07	192.932	191.131	184.504	184.042	180.876
Nov. 2008	180.56	191.127	187.827	190.361	183.577	182.122
Dec. 2008	189.03	189.916	184.999	202.980	183.471	184.610
Jan. 2009	182.07	191.995	185.613	195.753	184.210	185.233
Feb. 2009	167.35	189.940	184.397	167.580	184.338	185.274
Mar. 2009	189.30	183.988	178.988	185.936	183.725	184.247
Apr. 2009	175.84	189.348	181.395	180.165	184.529	184.930
MAPE (%)		6.044	4.636	3.799	3.638	3.371

Table 3.23 Wilcoxon signed-rank test

	Wilcoxon signed-rank test	
	$\alpha = 0.025$	$\alpha = 0.05$
Compared models	$W = 2$	$W = 3$
SVRCACO vs. ARIMA(1,1,1)	1[a]	1[a]
SVRCACO vs. GRNN($\sigma = 3.33$)	3	2[a]
SVRCACO vs. TF-ε-SVR-SA	2[a]	2[a]
SVRCACO vs. SVRPSO	2[a]	2[a]

[a]Denotes that SVRCACO model significantly outperforms other alternative models

models with other alternative models. The proposed SVRCACO model has smaller MAPE values than other alternative models. Furthermore, to verify the significance of accuracy improvement of SVRCACO model comparing with other alternative models, the Wilcoxon signed-rank test and asymptotic test are also conducted. The test results are shown in Tables 3.23 and 3.24, respectively. Clearly, the SVRCACO model is significantly superior to other alternative models, except GRNN model (minor significantly superior to GRNN model, only receives significance with $\alpha = 0.05$ level in Wilcoxon test, but all fails with both levels in asymptotic test) and SVRPSO model (not completely significantly outperforms SVRPSO model, only receives significance with both levels in Wilcoxon test, but all fails with both levels in asymptotic test). Figure 3.20 is provided to illustrate the forecasting accuracy among different models.

Table 3.24 Asymptotic test

Compared models	Asymptotic (S_1) test $\alpha = 0.05$	$\alpha = 0.10$
SVRCACO vs. ARIMA(1,1,1)	H_0: $e_1 = e_2$ $S_1 = -7.174$; $p = 0.000$ (reject H_0)	H_0: $e_1 = e_2$ $S_1 = -7.174$; $p = 0.000$ (reject H_0)
SVRCACO vs. GRNN($\sigma = 3.33$)	H_0: $e_1 = e_2$ $S_1 = -1.201$; $p = 0.1149$ (not reject H_0)	H_0: $e_1 = e_2$ $S_1 = -1.201$; $p = 0.1149$ (not reject H_0)
SVRCACO vs. TF-ε-SVR-SA	H_0: $e_1 = e_2$ $S_1 = -2.018$; $p = 0.0218$ (reject H_0)	H_0: $e_1 = e_2$ $S_1 = -2.018$; $p = 0.0218$ (reject H_0)
SVRCACO vs. SVRPSO	H_0: $e_1 = e_2$ $S_1 = 0.6341$; $p = 0.263$ (not reject H_0)	H_0: $e_1 = e_2$ $S_1 = 0.6341$; $p = 0.263$ (not reject H_0)

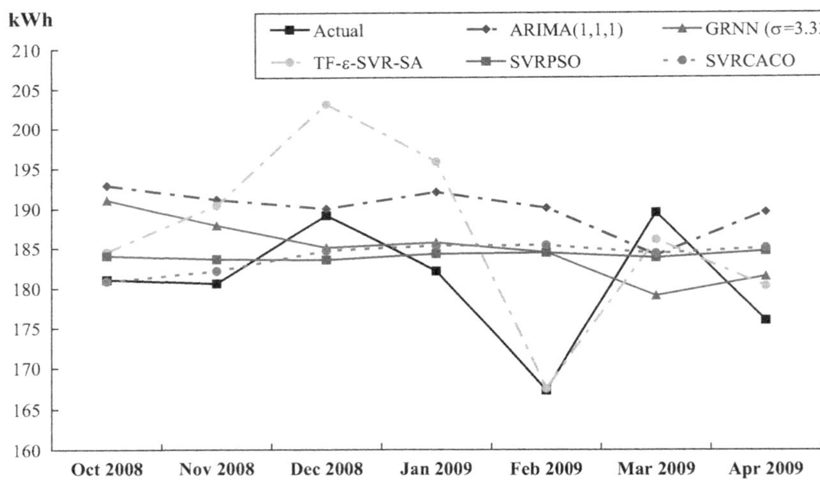

Fig. 3.20 Forecasting results of ARIMA, GRNN, TF-ε-SVR-SA, SVRPSO, and SVRCACO models

In this section, the CACO algorithm is also applied to overcome the shortcomings of GA and SA algorithms, that is, with memory to store the knowledge of good solutions and sharing information with each other. CACO algorithm has such the mechanism of learning/searching experiences storage and feedback to establish the shorter path (i.e., suitable parameter combination of a SVR model) between ant nest and food (i.e., smaller forecasting error). Therefore, it is expected to receive better forecasting performance than SVRGA and SVRSA models and has the potential to provide some competitive solution comparing with SVRPSO model. However, it is also clear that SVRCACO model is not fitting the actual electric loads very well even it has significant smaller MAPE values than other alternatives. Therefore, it also still requires hybridizing other novel techniques to improve this shortcoming.

3.8 Artificial Bee Colony Algorithm in SVR's Parameter Determination

3.8.1 Behaviors of Real Bees

The artificial bee colony (ABC) algorithm, proposed by Karaboga et al. [39] and further developed by Karaboga and Basturk [40–42], is inspired by the intelligent foraging behavior of honeybee swarm. As known that lots of optimization algorithms conduct only one search operation during one iteration time, for example, PSO algorithm carries out global search at the beginning and local search in the later stage. For ABC algorithm, it conducts both global search and local search during one iteration time to ensure that ABC algorithm is more probabilistic to receive more suitable parameter combination, and thereby efficiently to avoid local optimum to a large extent, and to receive better performance in optimization problem compared with GA, differential evolution (DE), and PSO algorithm [40–43].

The whole fabric of honeybee society depends on various communication ways among bees, such as waggle dance and special odor, to easily find food sources that produce relatively high amount of nectar [44]. To introduce this kind of forage selection model that leads to the emergence of collective intelligence of honeybee swarms, three essential components are defined: food sources, unemployed foragers, and employed foragers. In addition, two leading modes of the behavior, recruitment to a nectar source and abandonment of a source, are also embedded [42]:

1. Food sources (A and B in Fig. 3.21). The value of a food source depends on many factors, such as its proximity to the nest, richness or concentration of energy, and the ease of extracting this energy. For the sake of simplicity, the "profitability" of a food source can be represented with a single quantity.
2. Unemployed foragers (UF in Fig. 3.21). It is assumed that a bee has no prior knowledge about the food sources in the search field; thus, bee initializes its search as an unemployed forager. Unemployed foragers are looking for a food source to exploit. There are three types of unemployed foragers: scouts, onlookers, and recruits.

   (a) Scouts (S in Fig. 3.21). Without any prior knowledge, they start searching spontaneously for new food sources around the nest. The mean number of scouts averaged over conditions is about 10 % in nature [45].

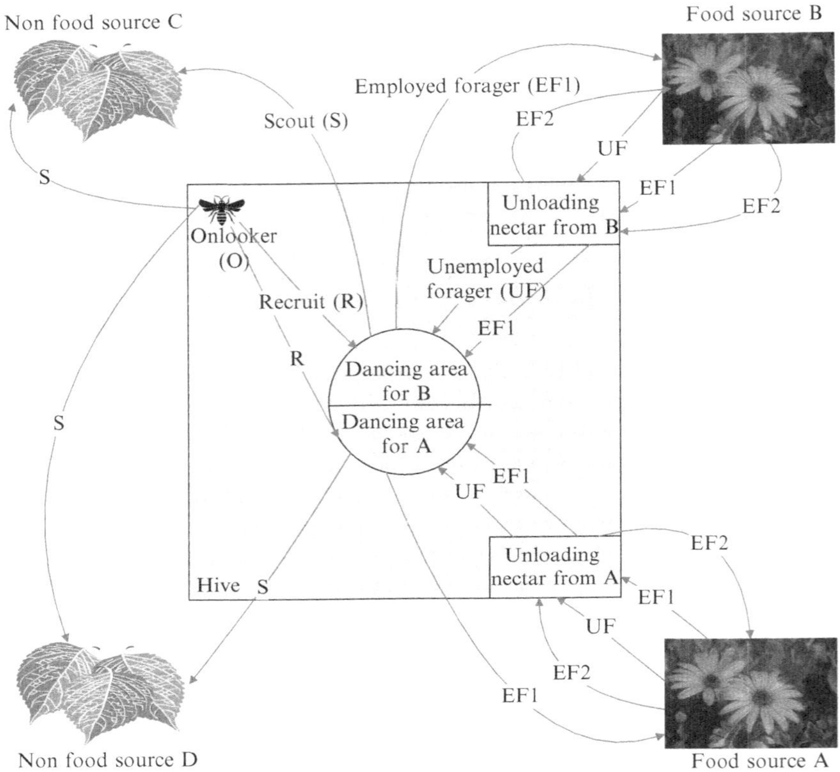

Fig. 3.21 Behavior of honeybee foraging for nectar

(b) Onlookers (O in Fig. 3.21). They wait in the nest and find a food source through the shared information from the employed foragers. There is a greater probability of onlookers choosing more profitable sources [43].

(c) Recruits (R in Fig. 3.21). If the onlookers attend a waggle dance done by some other bees, these onlookers will become recruits and start searching by using the obtained (shared) knowledge from the waggle dance.

3. Employed foragers. They are associated with a particular food source which they are currently exploiting or are "employed" at. They carry the information (profitability) about this particular source and share this information by a certain probability. After the employed foraging bee loads a portion of nectar from the food source, it returns to the hive and unloads the nectar to the food area in the hive. Then, the foraging bee has three possible behaviors related to residual amount of nectar to be acted.

   (a) Unemployed foragers (UF in Fig. 3.21). If the nectar amount is decreased to a low critical level or exhausted, the foraging bee abandons the food source and becomes an unemployed bee.

(b) Employed foragers type 1 (EF1 in Fig. 3.21). The foraging bee can go to the dance area to perform waggle dance to inform the nest mates about the food source.
(c) Employed foragers type 2 (EF2 in Fig. 3.21). On the contrary, if there are still sufficient amounts of nectar in the food source, the foraging bee can continue to forage without communicating the food source information with the nest mates.

The communication of information among bees is the most important matter in the whole fabric of honeybee society. Communication among bees related to the quality of food sources occurs in the dancing area. The related dance is called waggle dance, which closely provides the information correlated with the direction of and the distance from the food sources. Employed foragers share their information with a probability, which is proportional to the profitability of the food source. Hence, the recruitment is proportional to the profitability of a food source [46].

3.8.2 Operation Procedure of ABC Algorithm

The proposed procedure of ABC algorithm is illustrated as follows and the flowchart is shown as Fig. 3.22:

Step 1: Initialization. Initialize the population size N_p, the number of employed foragers n_e, and the number of unemployed foragers (onlookers) n_0, which satisfy the condition, $N_p = n_e + n_0$. Let x_{ij} ($i = 1, 2, \ldots, N_p$; $j = 1, 2, \ldots, D$) represent the initial solution of parameter combination in a SVR model, where D is the number of parameters. D is set as 3 in this book.

Step 2: Criteria of food source determination. Based on ABC algorithm, choosing a food source of an onlooker is dependent on the probability value associated with that food source. However, for forecasting accuracy improvement in the investigation, the onlooker will choose a food source according to the mean absolute percentage error (MAPE), shown as Eq. (3.13).

Step 3: Generate neighboring food sources (potential solutions). A candidate food source (solution) v_{ij} from the old solution x_{ij} can be generated as Eq. (3.30):

$$v_{ij} = x_{ij} + \Phi_{ij}(x_{ij} - x_{kj}), \tag{3.30}$$

where $k \in \{1, 2, \ldots, N\}$ is randomly chosen index, k must be different from i, and Φ_{ij} is a random number in the range $[-1,1]$. If the MAPE value of the candidate solution v_{ij} is equal to or smaller than x_{ij}'s, then v_{ij} will be set as the new solution; else, x_{ij} will remain as the solution. The parameter Φ_{ij} in ABC is the key factor that affects convergence [65].

Step 4: Determine the abandoned food source. If a solution cannot be improved through a predetermined threshold (limited iterations), then the food source is

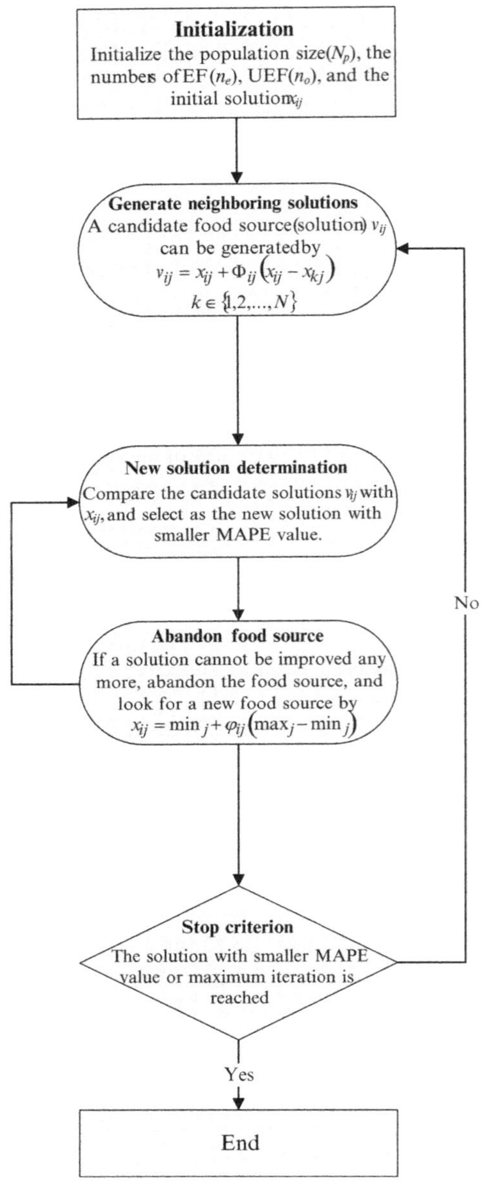

Fig. 3.22 The architecture of artificial bee colony (ABC) algorithm

considered to be abandoned. The employed bee will reinstate to be a scout, according to Eq. (3.31), to look for another new food source to replace the abandoned source:

$$x_{ij} = \min_j + \varphi_{ij}(\max_j - \min_j), \tag{3.31}$$

Table 3.25 Parameter determination of SVRABC model

Nos. of fed-in data	Parameters			MAPE of testing (%)
	σ	C	ε	
5	115.78	130.01	2.9542	3.812
10	193.26	44.08	2.4476	3.665
15	30.27	9,652.50	12.7640	3.509
20	620.15	4,246.00	13.1820	3.588
25	**38.348**	**4,552.10**	**16.8450**	**3.458**

where $\max_j$ is the maximal solution, that is, $\max_j = \max\{x_{1j}, x_{2j}, \ldots, x_{Nj}\}$; $\min_j$ represents the minimal solution, that is, $\min_j = \min\{x_{1j}, x_{2j}, \ldots, x_{Nj}\}$; and φ_{ij} is a random number in the range [−1,1].

Step 5: Stop criterion. If the new food source value is with smaller MAPE value or maximum iteration is reached, then the new three parameters $x_i^{(n+1)}$ and its corresponding objective value are the final solution; otherwise, go to the next iteration and go back to step 2.

3.8.3 ABC Algorithm for Three-Parameter Determination and Forecasting Results

This subsection will demonstrate the three-parameter determination of the proposed hybrid model (SVR with ABC), namely, SVRABC model.

Similarly, in the training stage, the rolling-based forecasting procedure is also employed to obtain the forecasting load and receive training error in the training stage. Then, if training error improvement occurs, the three kernel parameters, σ, C, and ε, of the SVRABC model adjusted by ABC algorithm are employed to calculate the validation error. The adjusted parameters with minimum validation error are also selected as the most appropriate parameters. The forecasting results and the suitable parameters for the SVRABC model are illustrated in Table 3.25, in which it is also indicated that these two models all perform the best when 25 fed-in data are used.

For simplified comparison among alternative models, SVRGA, SVRSA, and SVRGASA models are not considered due to their law forecasting accuracy levels; SVRCACO model is also not included in comparison due to minor relationship between CACO and ABC algorithms. Table 3.26 shows the actual values and the forecast values obtained using various forecasting models: ARIMA(1,1,1), GRNN ($\sigma = 3.33$), TF-ε-SVR-SA, SVRPSO, and SVRABC. The MAPE values are calculated to compare fairly the proposed models with other alternative models. The proposed SVRABC model has smaller MAPE values than other alternative models. Furthermore, to verify the significance of accuracy improvement of SVRABC model comparing with other alternative models, the Wilcoxon signed-rank test and asymptotic test are also conducted. The test results are shown in Tables 3.27

Table 3.26 Forecasting results of ARIMA, GRNN, TF-ε-SVR-SA, SVRPSO, and SVRABC models (unit, hundred million kWh)

Time point (month)	Actual	ARIMA (1,1,1)	GRNN ($\sigma = 3.33$)	TF-ε-SVR-SA	SVRPSO	SVRABC
Oct. 2008	181.07	192.932	191.131	184.504	184.042	184.498
Nov. 2008	180.56	191.127	187.827	190.361	183.577	183.372
Dec. 2008	189.03	189.916	184.999	202.980	183.471	183.323
Jan. 2009	182.07	191.995	185.613	195.753	184.210	183.549
Feb. 2009	167.35	189.940	184.397	167.580	184.338	183.774
Mar. 2009	189.30	183.988	178.988	185.936	183.725	183.999
Apr. 2009	175.84	189.348	181.395	180.165	184.529	183.420
MAPE (%)		6.044	4.636	3.799	3.638	3.458

Table 3.27 Wilcoxon signed-rank test

	Wilcoxon signed-rank test	
	$\alpha = 0.025$	$\alpha = 0.05$
Compared models	$W = 2$	$W = 3$
SVRABC vs. ARIMA(1,1,1)	1[a]	1[a]
SVRABC vs. GRNN($\sigma = 3.33$)	2[a]	2[a]
SVRABC vs. TF-ε-SVR-SA	2[a]	2[a]
SVRABC vs. SVRPSO	2[a]	2[a]

[a]Denotes that SVRABC model significantly outperforms other alternative models

Table 3.28 Asymptotic test

	Asymptotic (S_1) test	
Compared models	$\alpha = 0.05$	$\alpha = 0.10$
SVRABC vs. ARIMA (1,1,1)	H_0: $e_1 = e_2$ $S_1 = -13.231$; $p = 0.000$(reject H_0)	H_0: $e_1 = e_2$ $S_1 = -13.231$; $p = 0.000$(reject H_0)
SVRABC vs. GRNN ($\sigma = 3.33$)	H_0: $e_1 = e_2$ $S_1 = -2.257$; $p = 0.01199$ (reject H_0)	H_0: $e_1 = e_2$ $S_1 = -2.257$; $p = 0.01199$ (reject H_0)
SVRABC vs. TF-ε-SVR-SA	H_0: $e_1 = e_2$ $S_1 = -2.066$; $p = 0.0194$(reject H_0)	H_0: $e_1 = e_2$ $S_1 = -2.066$; $p = 0.0194$(reject H_0)
SVRABC vs. SVRPSO	H_0: $e_1 = e_2$ $S_1 = -2.723$; $p = 0.0032$(reject H_0)	H_0: $e_1 = e_2$ $S_1 = -2.723$; $p = 0.0032$(reject H_0)

and 3.28, respectively. Clearly, the SVRABC model is significantly superior to other alternative models. Figure 3.23 is provided to illustrate the forecasting accuracy among different models.

In this section, the ABC algorithm is also applied to overcome the shortcomings of PSO algorithm (only carries out global search at the beginning and local search in the later stage), that is,, conducting both global search and local search during one

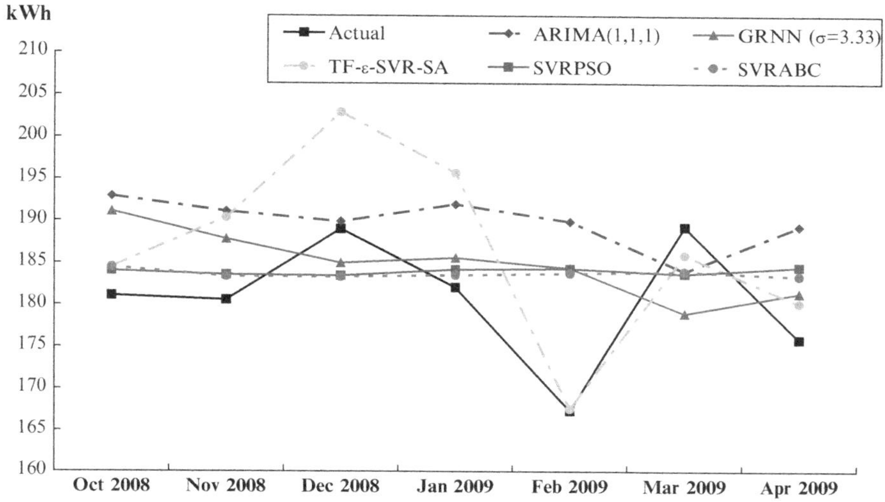

Fig. 3.23 Forecasting results of ARIMA, GRNN, TF-ε-SVR-SA, SVRPSO, and SVRABC models

iteration time to enrich the searching behavior to avoid trapping into local optimum. Therefore, it is also expected to receive better forecasting performance than SVRPSO model and has the potential to provide some alternative solution comparing with SVRCACO model. However, it is also clear that SVRABC model is not fitting the actual electric loads very well even if it has significant smaller MAPE values than other alternatives. Therefore, it also still requires hybridizing other novel techniques to improve this shortcoming.

3.9 Immune Algorithm in SVR's Parameter Determination

3.9.1 Operation Procedure of IA

The immune algorithm (IA), proposed by Mori et al. [47] and used in this book, is based on the learning mechanism of natural immune systems. Similar to GA, SA, and PSO, IA is also a population-based evolutionary algorithm; therefore, it provides a set of solutions for exploration and exploitation of search space to obtain optimal/near optimal solutions [48]. The natural immune system is a complex adaptive system that efficiently employs several mechanisms to recognize all cells within the body and classify those cells as self or nonself. Additionally, the nonself cells are further categorized to stimulate an appropriate type of defensive mechanism for defending against foreign invaders, such as bacteria and viruses. The lymphocyte is the main type of immune cell participating in the immune response. The lymphocyte contains two subclasses: *T* and *B*. Each subclass has its own function. When an antigen enters

Fig. 3.24 The architecture of immune algorithm (IA)

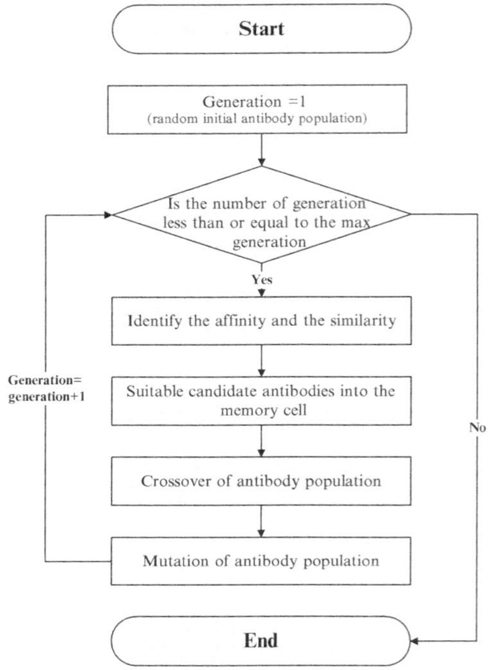

the bloodstream and lymphatic system, the antigen encounters B-cells, and antibodies anchored in the membrane of B-cells recognize antigens in the bacteria. T-cells, which have already received communication from macrophages about the antigen, then communicate with B-cells and stimulate their proliferation. The proliferated B-cells turn into memory cells and produce antibodies. After the antibodies enter the bloodstream via the heart, the antibodies bind to antigens and kill them with the help of macrophages and other proteins.

Analogous to the natural immune system, the IA has the ability to seek out the best solution for optimization problems. In the IA procedure, the optimization problem can be viewed as antigens. Conversely, the feasible solutions of the optimization problem are treated as antibodies (B-cells). The procedure of IA is illustrated as follows and the flowchart is shown as Fig. 3.24.

Step 1: Random initialization of antibody population. The initial antibody population represented by binary-code string, including three parameters (σ, C, and ε) of a SVR model, is generated randomly. For example, assume that an antibody contains 12 binary codes to represent three SVR parameters. Each parameter is thus expressed by four binary codes. Therefore, for example, assume the set boundaries for parameters σ, C, and ε are 2, 10, and 0.5, respectively; then, the antibody with binary code "1 0 0 1 0 1 0 1 0 0 1 1" implies that the real values of the three parameters σ, C, and ε are 1.125, 3.125, and 0.09375, respectively. The number of initial antibodies is the same as the size of the memory cell. The size of the memory cell is set to 10 in this book.

Step 2: Identifying the affinity and the similarity. A higher affinity value implies that an antibody has a higher activation with an antigen. To maintain the diversity of the antibodies stored in the memory cells, the antibodies with lower similarity have higher probability of being included in the memory cell. Therefore, an antibody with a higher affinity value and a lower similarity value has a good likelihood of entering the memory cells. The affinity between the antibody and antigen is defined as Eq. (3.32):

$$\mathrm{Ag}_k = 1/(1 + d_k), \tag{3.32}$$

where d_k denotes the SVR forecasting errors obtained by the antibody k.

The similarity between antibodies is expressed as Eq. (3.33):

$$\mathrm{Ab}_{ij} = 1/(1 + T_{ij}), \tag{3.33}$$

where T_{ij} denotes the difference between the two SVR forecasting errors obtained by the antibodies inside (existed) and outside (will be entering) the memory cell.

Step 3: Selection of antibodies in the memory cell. Antibodies with higher values of Ag_k are considered to be potential candidates for entering the memory cell. However, the potential antibody candidates with Ab_{ij} values exceeding a certain threshold are not qualified to enter the memory cell. In this investigation, the threshold value is set to 0.9.

Step 4: Crossover and mutation of antibody population. New antibodies are created via crossover and mutation operations. To perform crossovers, strings representing antibodies are paired randomly. Moreover, segments of paired strings between two determined breakpoints are swapped. Mutations are performed randomly by converting a "1" code into a "0" code or a "0" code in to a "1" code. The crossover and mutation rates are determined using probabilities. In this investigation, the probabilities are set to 0.5 and 0.1 for crossover and mutation, respectively.

Step 5: Stopping criterion. If the number of generations equals a given scale, then the best antibody is a solution; otherwise, return to step 2.

The IA is used to seek a better combination of the three parameters in SVR. The value of the mean absolute percent error (MAPE) is used as the criterion (the smallest value of MAPE) of forecasting errors to determine the suitable parameters used in SVR model, which is given by Eq. (3.13).

3.9.2 IA for Three-Parameter Determination and Forecasting Results

This subsection will demonstrate the three-parameter determination of the proposed hybrid model (SVR with IA), namely, SVRIA model.

Similarly, in the training stage, the rolling-based forecasting procedure is also employed to obtain the forecasting load and receive training error in the training

Table 3.29 Parameter determination of SVRIA model

Nos. of fed-in data	Parameters			MAPE of testing (%)
	σ	C	ε	
5	758.12	409.33	3.7736	4.940
10	11.74	180.91	0.6728	4.079
15	43.21	2,367.70	13.5250	3.504
20	282.38	2,365.50	2.4397	3.880
25	**149.93**	**4,293.10**	**9.4790**	**3.211**

Table 3.30 Forecasting results of ARIMA, GRNN, TF-ε-SVR-SA, SVRCACO, SVRABC, and SVRIA models (unit, hundred million kWh)

Time point (month)	Actual	ARIMA (1,1,1)	GRNN ($\sigma = 3.33$)	TF-ε-SVR-SA	SVRCACO	SVRABC	SVRIA
Oct. 2008	181.07	192.932	191.131	184.504	180.876	184.498	181.322
Nov. 2008	180.56	191.127	187.827	190.361	182.122	183.372	181.669
Dec. 2008	189.03	189.916	184.999	202.980	184.610	183.323	183.430
Jan. 2009	182.07	191.995	185.613	195.753	185.233	183.549	183.964
Feb. 2009	167.35	189.940	184.397	167.580	185.274	183.774	184.030
Mar. 2009	189.30	183.988	178.988	185.936	184.247	183.999	182.829
Apr. 2009	175.84	189.348	181.395	180.165	184.930	183.420	183.463
MAPE (%)		6.044	4.636	3.799	3.371	3.458	3.211

stage. Then, if training error improvement occurs, the three kernel parameters, σ, C, and ε, of the SVRIA model adjusted by IA are employed to calculate the validation error. The adjusted parameters with minimum validation error are also selected as the most appropriate parameters. The forecasting results and the suitable parameters for the SVRIA model are illustrated in Table 3.29, in which it is also indicated that these two models all perform the best when 25 fed-in data are used.

For simplified comparison among alternative models, SVRGA, SVRSA, SVRGASA, and SVRPSO models are not considered due to their law forecasting accuracy levels. Table 3.30 shows the actual values and the forecast values obtained using various forecasting models: ARIMA(1,1,1), GRNN($\sigma = 3.33$), TF-ε-SVR-SA, SVRCACO, and SVRABC. The MAPE values are calculated to compare fairly the proposed models with other alternative models. The proposed SVRIA model has smaller MAPE values than other alternative models. Furthermore, to verify the significance of accuracy improvement of SVRIA model comparing with other alternative models, the Wilcoxon signed-rank test and asymptotic test are also conducted. The test results are shown in Tables 3.31 and 3.32, respectively. Clearly, the SVRIA model is almost significantly superior to other alternative models, except SVRABC model (only receives significance with $\alpha = 0.05$ level in Wilcoxon test, but all fails with both levels in asymptotic test). Figure 3.25 is provided to illustrate the forecasting accuracy among different models.

Table 3.31 Wilcoxon signed-rank test

	Wilcoxon signed-rank test	
	$\alpha = 0.025$	$\alpha = 0.05$
Compared models	$W = 2$	$W = 3$
SVRIA vs. ARIMA(1,1,1)	0[a]	0[a]
SVRIA vs. GRNN($\sigma = 3.33$)	2[a]	2[a]
SVRIA vs. TF-ε-SVR-SA	2[a]	2[a]
SVRIA vs. SVRCACO	1[a]	1[a]
SVRIA vs. SVRABC	3	3[a]

[a]Denotes that SVRIA model significantly outperforms other alternative models

Table 3.32 Asymptotic test

	Asymptotic (S_1) test	
Compared models	$\alpha = 0.05$	$\alpha = 0.10$
SVRIA vs. ARIMA(1,1,1)	H_0: $e_1 = e_2$ $S_1 = -9.143$; $p = 0.000$ (reject H_0)	H_0: $e_1 = e_2$ $S_1 = -9.143$; $p = 0.000$ (reject H_0)
SVRIA vs. GRNN($\sigma = 3.33$)	H_0: $e_1 = e_2$ $S_1 = -1.768$; $p = 0.03856$ (reject H_0)	H_0: $e_1 = e_2$ $S_1 = -1.768$; $p = 0.03856$ (reject H_0)
SVRIA vs. TF-ε-SVR-SA	H_0: $e_1 = e_2$ $S_1 = -3.910$; $p = 0.000$ (reject H_0)	H_0: $e_1 = e_2$ $S_1 = -3.910$; $p = 0.000$ (reject H_0)
SVRIA vs. SVRCACO	H_0: $e_1 = e_2$ $S_1 = -3.632$; $p = 0.00014$ (reject H_0)	H_0: $e_1 = e_2$ $S_1 = -3.632$; $p = 0.00014$ (reject H_0)
SVRIA vs. SVRABC	H_0: $e_1 = e_2$ $S_1 = 0.218$; $p = 0.4136$ (not reject H_0)	H_0: $e_1 = e_2$ $S_1 = 0.218$; $p = 0.4136$ (not reject H_0)

In this section, the IA is also employed to overcome the shortcomings of GA, SA, and PSO algorithms, that is, providing a set of solutions for exploration and exploitation of search space to obtain optimal/near optimal solutions by using immune system to find out the feasible solutions of the optimization problem. IA has such a mechanism to recognize all cells within the body and classify those cells as self or nonself; the nonself cells are categorized to the defensive mechanism for defending against foreign invaders; the lymphatic system contains two subclasses, T-cells and B-cells, to communicate with each other when an antigen enters the bloodstream. The proliferated B-cells turn into memory cells and produce antibodies (i.e., suitable parameter combination of a SVR model). The optimization problem (i.e., smaller forecasting error) is viewed as antigens. Conversely, the feasible solutions of the optimization problem are treated as antibodies (B-cells). Therefore, it is expected to receive better forecasting performance than SVRGA, SVRSA, and SVRPSO models and has the potential to provide some competitive

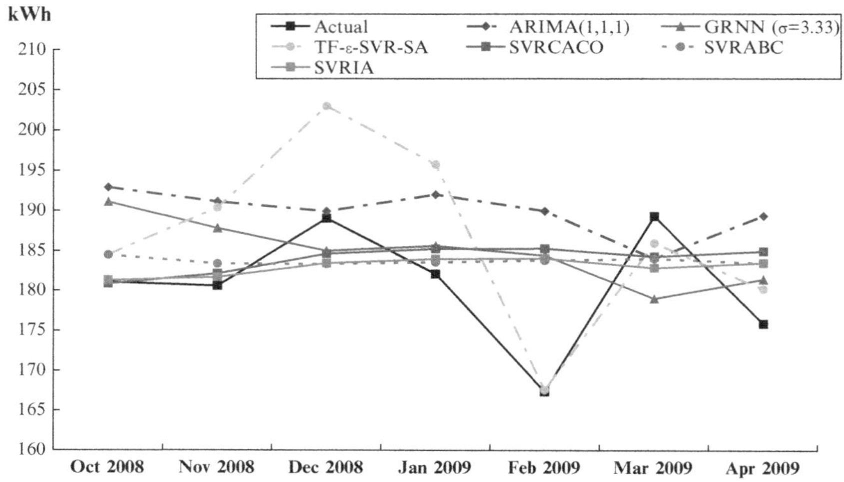

Fig. 3.25 Forecasting results of ARIMA, GRNN, TF-ε-SVR-SA, SVRCACO, SVRABC, and SVRIA models

solution comparing with SVRCACO and SVRABC models. However, it is also clear that SVRIA is not fitting the actual electric loads very well even if it has significant smaller MAPE values than other alternatives. Therefore, it also still requires hybridizing other novel techniques to improve this shortcoming.

References

1. Wang J, Zhu W, Zhang W, Sun D (2009) A trend fixed on firstly and seasonal adjustment model combined with the ε-SVR for short-term forecasting of electricity demand. Energ Policy 37:4901–4909. doi:10.1016/j.enpol.2009.06.046
2. Daniel WW (1978) Applied nonparametric statistics. Houghton Mifflin Co., Boston, MA
3. Diebold FX, Mariano RS (1995) Comparing predictive accuracy. J Bus Econ Stat 13:253–263. doi:10.1080/07350015.1995.10524599
4. Morgan WA (1939) A test for the significance of the difference between the two variances in a sample from a normal bivariate population. Biometrika 31:13–19. doi:10.1093/biomet/31.1-2.13
5. Granger CWJ, Newbold P (1977) Forecasting economic time series. Academic, Orlando, FL
6. Meese RA, Rogoff K (1988) Was it real? the exchange rate—interest differential relation over the modern floating-rate period. J Financ 43:933–948. doi:10.1111/j.1540-6261.1988.tb02613.x
7. Holland J (1975) Adaptation in natural and artificial system. University of Michigan Press, Ann Arbor, MI
8. Cercignani C (1988) The Boltzmann equation and its applications. Springer, Berlin
9. Metropolis N, Rosenbluth AW, Rosenbluth MN, Teller AH (1953) Equations of state calculations by fast computing machines. J Chem Phys 21:1087–1092. doi:10.1063/1.1699114
10. Kirkpatrick S, Gelatt CD, Vecchi MP (1983) Optimization by simulated annealing. Science 220:671–680. doi:10.1126/science.220.4598.671
11. Van Laarhoven PJM, Aarts EHL (1987) Simulated annealing: theory and applications. Kluwer Academic, Dordrecht

12. Dekkers A, Aarts EHL (1991) Global optimization and simulated annealing. Math Program 50:367–393. doi:10.1007/BF01594945
13. Lee J, Johnson GE (1983) Optimal tolerance allotment using a genetic algorithm and truncated Monte Carlo simulation. Comput Aided Des 25:601–611. doi:10.1016/0010-4485(93)90075-Y
14. Shieh HJ, Peralta RC (2005) Optimal in situ bioremediation design by hybrid genetic algorithm-simulated annealing. J Water Resour Plan Manage-ASCE 131:67–78. doi:10.1061/(ASCE) 0733-9496(2005) 131:1(67)
15. Ponnambalam SG, Reddy MM (2003) A GA-SA multiobjective hybrid search algorithm for integrating lot sizing and sequencing in flow-line scheduling. Int J Adv Manuf Technol 21:126–137. doi:10.1007/s001700300015
16. Zhao F, Zeng X (2006) Simulated annealing—genetic algorithm for transit network optimization. J Comput Civil Eng 20:57–68. doi:10.1061/(ASCE)0887- 3801(2006) 20:1(57)
17. Cordón O, Moya F, Zarco C (2002) A new evolutionary algorithm combining simulated annealing and genetic programming for relevance feedback in fuzzy information retrieval systems. Soft Comput 6:308–319. doi:10.1007/s00500- 002-0184-8
18. Ganesh K, Punniyamoorthy M (2005) Optimization of continuous-time production planning using hybrid genetic algorithms-simulated annealing. Int J Adv Manuf Technol 26:148–154. doi:10.1007/s00170-003-1976-4
19. Wang ZG, Wong YS, Rahman M (2004) Optimisation of multi-pass milling using genetic algorithm and genetic simulated annealing. Int J Adv Manuf Technol 24:727–732. doi:10.1007/s00170-003-1789-5
20. Bergey PK, Ragsdale CT, Hoskote M (2003) A simulated annealing genetic algorithm for the electrical power districting problem. Ann Oper Res 121:33–55. doi:10.1023/A:1023347000978
21. Kennedy J, Eberhart RC (1995) Particle swarm optimization. In: Proceedings of IEEE international conference neural networks, Washington, DC, pp 1942–1948. doi:10.1109/ICNN.1995.488968
22. Eberhart RC, Shi Y (2001) Particle swarm optimization: developments, applications and resources. In: Proceedings of the 2001 congress on evolutionary computation, Seoul, South Korea, pp 81–86. doi:10.1109/CEC.2001.934374
23. Angeline PJ (1998) Evolutionary optimization versus particle swarm optimization: philosophy and performance differences. In: Proceedings of the 7th international conference on evolutionary programming, San Diego, CA, pp 601–610. doi:10.1007/BFb0040811
24. Liu B, Wang L, Jin YH, Tang F, Huang DX (2005) Improved particle swarm optimization combined with chaos. Chaos Soliton Fract 25:1261–1271. doi:10.1016/j.chaos.2004.11.095
25. Dorigo M (1992) Optimization, learning, and natural algorithms (Doctoral Dissertation), Dipartimento di Elettronica e Informazione, Politecnico di Milano, Italy
26. Dorigo M, Maniezzo V, Colorni A (1996) Ant system: optimization by a colony of cooperating ants. IEEE Trans Syst Man Cybern B Cybern 26:29–41. doi:10.1109/3477.484436
27. Colorni A, Dorigo M, Maniezzo V, Trubian M (1994) Ant system for job-shop scheduling. Belg J Oper Res Stat Comput Sci 34:39–53
28. Dorigo M, Gambardella L (1997) Ant colony system: a cooperative learning approach to the traveling salesman problem. IEEE Trans Evol Comput 1:53–66. doi:10.1109/4235.585892
29. Bland JA (1999) Space-planning by ant colony optimization. Int J Comput Appl Technol 12:320–328. doi:10.1504/IJCAT.1999.000215
30. Maniezzo V, Colorni A (1999) The ant system applied to the quadratic assignment problem. IEEE Trans Knowl Data Eng 11:769–778. doi:10.1109/69.806935
31. Parpinelli RS, Lopes HS, Freitas AA (2002) Data mining with an ant colony optimization algorithm. IEEE Trans Evol Comput 6:321–332. doi:10.1109/TEVC. 2002.802452
32. Bilchev G, Parmee IC (1995) The ant colony metaphor for searching continuous design spaces. Lect Notes Comput Sci 993:25–39. doi:10.1007/3-540-60469-3_ 22
33. Mathur M, Karale SB, Priye S, Jyaraman VK, Kulkarni BD (2000) Ant colony approach to continuous function optimization. Ind Eng Chem Res 39:3814–3822. doi:10.1021/ie990700g

34. Wodrich M, Bilchev G (1997) Cooperative distributed search: the ant's way. Control Cybern 26:413–446
35. Monmarche N, Venturini G, Slimane M (2000) On how Pachycondyla apicalis ants suggest a new search algorithm. Futur Gener Comp Syst 16:937–946. doi:10.1016/S0167-739X(00)00047-9
36. Dreo J, Siarry P (2002) A new ant colony algorithm using the heterarchical concept aimed at optimization of multiminima continuous functions. Lect Notes Comput Sci 2463:216–221
37. Socha K, Dorigo M (2008) Ant colony optimization for continuous domains. Eur J Oper Res 185:1155–1173. doi:10.1016/j.ejor.2006.06.046
38. Abbaspour KC, Schulin R, van Genuchten MT (2001) Estimating unsaturated soil hydraulic parameters using ant colony optimization. Adv Water Resour 24:827–841. doi:10.1016/S0309-1708(01)00018-5
39. Karaboga N, Kalinli A, Karaboga D (2004) Designing digital IIR filters using ant colony optimisation algorithm. Eng Appl Artif Intell 17(3):301–309. doi:10.1016/j.engappai.2004.02.009
40. Karaboga D, Basturk B (2007) A powerful and efficient algorithm for numerical function optimization: artificial bee colony (ABC) algorithm. J Global Optim 39(3):459–471. doi:10.1007/s10898-007-9149-x
41. Karaboga D, Basturk B (2007) Artificial bee colony (ABC) optimization algorithm for solving constrained optimization problems. Lect Notes Comput Sci 4529:789–798. doi:10.1007/978-3-540-72950-1_77
42. Karaboga D, Basturk B (2008) On the performance of artificial bee colony (ABC) algorithm. Appl Soft Comput 8:687–697. doi:10.1016/j.asoc.2007.05.007
43. Xu C, Duan H, Liu F (2010) Chaotic artificial bee colony approach to uninhabited combat air vehicle (UCAV) path planning. Aerosp Sci Technol 14:535–541. doi:10.1016/j.ast.2010.04.008
44. Fathian M, Amiri B, Maroosi A (2007) Application of honey bee mating optimization algorithm on clustering. Appl Math Comput 190:1502–1513. doi:10. 1016/j.amc.2007.02.029
45. Özbakir L, Baykasoglu A, Tapkan P (2010) Bees algorithm for generalized assignment problem. Appl Math Comput 215:3782–3795. doi:10.1016/j.amc. 2009.11.018
46. Tereshko V, Loengarov A (2005) Collective decision making in honey-bee foraging dynamics. Comput Inform Syst 9(3):1–7
47. Mori K, Tsukiyama M, Fukuda T (1993) Immune algorithm with searching diversity and its application to resource allocation problem. Trans Inst Electr Eng Jpn 113-C:872–878
48. Prakash A, Khilwani N, Tiwari MK, Cohen Y (2008) Modified immune algorithm for job selection and operation allocation problem in flexible manufacturing system. Adv Eng Softw 39:219–232. doi:10.1016/j.advengsoft. 2007.01.024

Chapter 4
Chaos/Cloud Theories to Avoid Trapping into Local Optimum

As demonstrated in Chap. 3, different evolutionary algorithms, including genetic algorithm (GA), simulated annealing (SA) algorithm, hybrid GASA algorithm, particle swarm optimization (PSO) algorithm, continuous ant colony optimization (CACO) algorithm, artificial bee colony (ABC) algorithm, and immune algorithm (IA), are employed to determine suitable parameter combination of an SVR-based electric load forecasting model. These forecasting results indicate that almost all SVR-based models with different evolutionary algorithms are superior to other competitive forecasting models (including ARIMA, HW, GRNN, and BPNN models); however, these algorithms almost lack knowledge memory or storage mechanisms which would be either time-consuming or inefficient in searching the suitable parameters, that is, premature convergence (being trapped in local optimum). For example, for SVRGA model, in GA processing, new individuals are generated by the following operators, selection, crossover, and mutation. For all types of objective functions, the generation begins with a binary coding for the parameter set. Based on this special binary coding process, GA is able to solve some specified problems which are not easily to be solved by traditional algorithms. GA can empirically provide a few best fitted offsprings from the whole population; however, after some generations, due to low diversity of the population, it might lead to a premature convergence [1]. For SVRSA model, SA algorithm is a generic probabilistic search technique that simulates the material physical process of heating and controlled cooling. Each step of SA attempts to replace the current state by a random move. The new state may then be accepted with a probability that depends both on the difference between the corresponding function values and also on a global parameter, temperature. Thus, SA has some institution to reach more ideal solutions. However, SA algorithm requires the subtle and skillful adjustment in the annealing schedule; therefore, the settings of the size of the temperature steps during annealing, the temperature range, the number of restarts, and the redirection of the search will be taken into account carefully [1]; in addition, its Monte Carlo scheme and lack of knowledge memory mechanism will also lead to time-consuming and inefficient searching in annealing process.

W.-C. Hong, *Intelligent Energy Demand Forecasting*, Lecture Notes in Energy 10, DOI 10.1007/978-1-4471-4968-2_4, © Springer-Verlag London 2013

To overcome these drawbacks of GA, it is necessary to find some effective approaches and improvements on GA to maintain the population diversity and avoid leading to misleading local optimum. Meanwhile, to overcome these shortcomings of SA algorithm, it is necessary to find some effective approaches and improvement on SA's subtle and skillful adjustment in the annealing schedule. One feasible improving approach is applying chaotic sequence to transform searching variables into chaotic variables, due to its easy implementation and unique ability to avoid becoming trapped in local optima [1]. Another feasible improving tool is the so-called cloud model, which can transform the uncertainty between quantitative representation and qualitative concept (in language form) [2], that is, it can successfully realize the transformation between a qualitative concept in words and numerical representation [2]. Therefore, it is suitable to be employed to solve the problem of temperature decreasing discretely in SA algorithm. This chapter introduces that hybrid chaotic sequence or cloud theory with evolutionary algorithms can overcome the shortcomings of trapping local optimum to improve forecasting performance.

4.1 Brief Introductions of Chaotic Sequence and Cloud Model

4.1.1 *Chaos and Mapping Functions*

Chaos is a ubiquitous phenomenon in the nonlinear system and is defined as highly unstable motion in finite phase space. It often occurs in deterministic nonlinear dynamic systems [3, 4]. Such motion is very similar to a random process ("randomicity"). Therefore, any variable in the chaotic space can travel ergodically over the whole space of interest ("ergodicity"). The variation of those chaotic variables has a delicate inherent rule, even though its variation looks like disorder ("regularity"). Additionally, chaotic behaviors are extremely sensitive to the initial condition, which is an important property, sometimes referred to as the so-called butterfly effect [5, 6]. Lorenz discovered that minute changes in initial conditions steered subsequent simulations toward radically different final stales when attempting to simulate a global weather system numerically. Based on the two advantages of the chaos, the chaotic optimization algorithm (COA) is proposed to solve complex function optimization [3]. The COA transforms the variable of problems from the solution space to chaos space and then performs a search to discover the solution by the three characteristics (*randomicity*, *ergodicity*, and *regularity*) of the chaotic variables. Therefore, chaotic variable may be adopted by utilizing these characteristics of chaotic phenomenon for global search and optimizing to increase the particle diversity.

Due to easy implementation process and special mechanism to escape from local optimum [1], chaos and chaos-based searching algorithms have received intense attention [7, 8]. Any decision variable in optimization problem can be chaotized by the chaotic sequence as a chaotic variable to carefully expand its searching space, that is, let variable travel ergodically over the searching space. The critical factor influencing the performance improvement is the chaotic mapping function. There are several mostly adopting chaotic mapping functions, such as the logistic mapping function [9] [Eq. (4.1)], the tent mapping function [10] [Eq. (4.2)], the An mapping function [11] [Eq. (4.3)], and the cat mapping function [12] [Eq. (4.4)] as the chaotic sequence generator.

$$x_{n+1} = \mu \cdot x_n(1 - x_n), \tag{4.1}$$

where x_n is the iteration value of the variable x in the nth time and μ is a control parameter; when $\mu = 4$, the system will be completely in a chaos state, and x_0 may take any initial value in (0,1) except 0.25, 0.5, and 0.75.

$$x_{n+1} = \begin{cases} 2x_n & x \in [0,0.5] \\ 2(1 - x_n) & x \in (0.5,1], \end{cases} \tag{4.2}$$

where x_n is the iteration value of the variable x in the nth time and n is the iteration times.

$$x_{n+1} = \begin{cases} \frac{3}{2}x_n + \frac{1}{4} & x \in [0,0.5) \\ \frac{1}{2}x_n - \frac{1}{4} & x \in [0.5,1], \end{cases} \tag{4.3}$$

where x_n is the iteration value of the variable x in the nth time and n is the iteration times.

$$\begin{cases} x_{n+1} = (x_n + y_n) \bmod 1 \\ y_{n+1} = (x_n + 2y_n) \bmod 1, \end{cases} \tag{4.4}$$

where $x \bmod 1 = x - [x]$.

To analyze the chaotic characteristics of these four mapping functions, set the initial value of these four mapping functions as 0.1, respectively, and set the iteration times as 50,000; the occurrence number of the obtained chaotic variable for each mapping function is recorded. The statistical results for each mapping function are as shown in Fig. 4.1, in which the value of the probability density for the chaotic sequence generated by the logistic mapping function is mostly distributed at both ends due to complying with Chebyshev distribution of more at both ends and less in middle; the chaotic sequence generated by the tent mapping function is affected by limited word length and limited precision of the computer and is rapid to fall into a minor cycle or a fixed point; the number of variables generated by the An mapping function gradually reduces along with variable values changing from small to large; however, different from above three mapping functions, the distribution of the cat mapping function is relatively uniform in the interval [0,1] and has no cyclic phenomenon during its iteration process. Therefore,

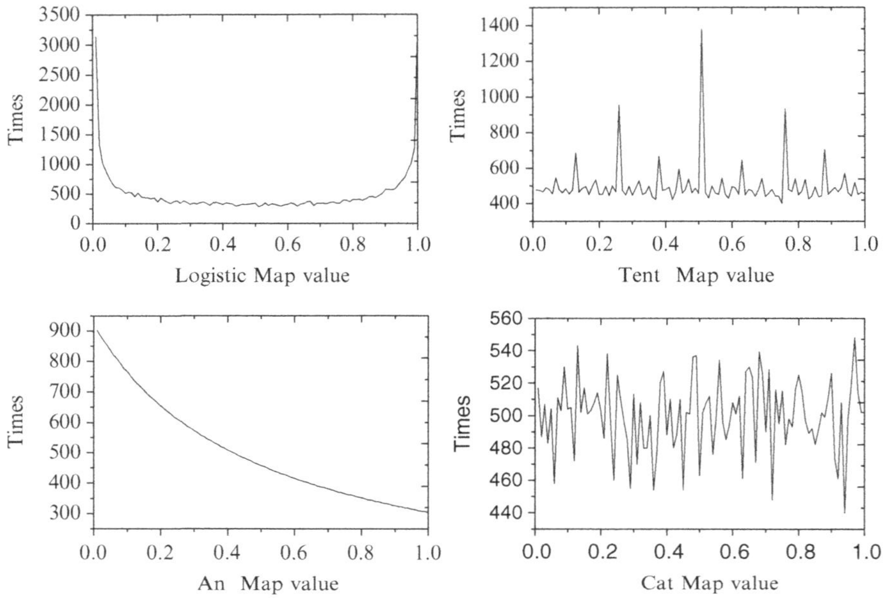

Fig. 4.1 Iterative distribution of four mapping functions

the cat mapping function has better chaotic distribution characteristic than other alternatives. The electric load reveals a univariate time series; thus, even the cat mapping function has better chaotic distribution characteristic; it is difficult to conduct two-dimensional transformation of these three parameters in an SVR model. Therefore, this book determines to employ logistic mapping function to transform three parameters into chaotic variables, with second good ergodic uniformity, and not easily to fall into minor cycle.

4.1.2 Chaotic Sequence by Logistic Mapping Function

As mentioned, chaos is an irregular nonlinear phenomenon in the natural world and is defined as the highly unstable unpredictable motion of deterministic systems in finite phase space. Thus, a nonlinear system is said to be chaotic if it exhibits sensitive dependence on initial conditions and has an infinite number of different periodic responses. This sensitive dependence on initial conditions is generally exhibited by systems containing multiple elements with nonlinear interactions. In addition, it is observed in not only complex systems but even in very simple logistic equations.

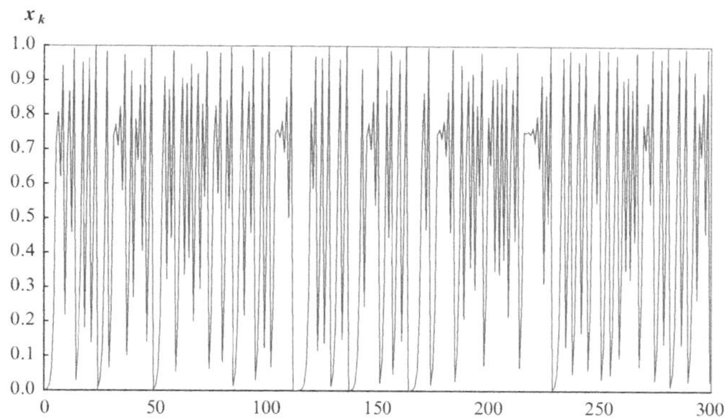

Fig. 4.2 Diversity of the logistic mapping function

Chaotic sequence can be represented by the famous logistic function (one-dimension base) defined by [9], as Eq. (4.1) and specified in Eq. (4.5) for hybridizing with the SVR mode.

$$x^{(i+1)} = \mu x^{(i)}(1 - x^{(i)}) \tag{4.5}$$

where $x^{(i)} \in (0,1)$, $i = 0, 1, 2, \ldots$ $x^{(i)}$ denotes the value of the chaotic variable x at the ith iteration, and μ denotes the so-called bifurcation parameter of the system, $\mu \in [0,4]$. The system behavior varies significantly with μ. The value of μ determines whether x stabilizes at a constant size, fluctuates between limited sequences of sizes, or behaves chaotically in an unpredictable pattern. For certain values of the parameter μ, such as $\mu = 4$, and $x^{(0)} \notin \{0.25,0.5,0.75\}$, this system exhibits chaotic behavior. Figure 4.2 shows the chaotic dynamics of the system where $x^{(0)} = 0.001$, $i = 300$. Clearly, a very small difference in the initial value of x causes a large difference in its future behavior, which is the basic characteristic of chaos. Additionally, x can travel ergodically over the whole space of interest; the variation of x has a delicate inherent rule, even though it looks like disorder.

4.1.3 Basic Concept of Cloud Model

Proposed by Li et al. [2], the cloud model is an uncertainty conversion model between qualitative knowledge description and quantitative value expression. The cloud model has the characteristics of uncertainty with certainty and stability with variation in knowledge expression, which embodies the basic principle of evolution of natural species. The cloud model has been successfully applied to such fields as intelligent control [13] and data mining [14]. Recently, the cloud model has also been introduced to the PSO algorithm, for example, Wei et al. [15] and Liu and Zhou [16] adjust the inertia weight of the particle with the cloud model and

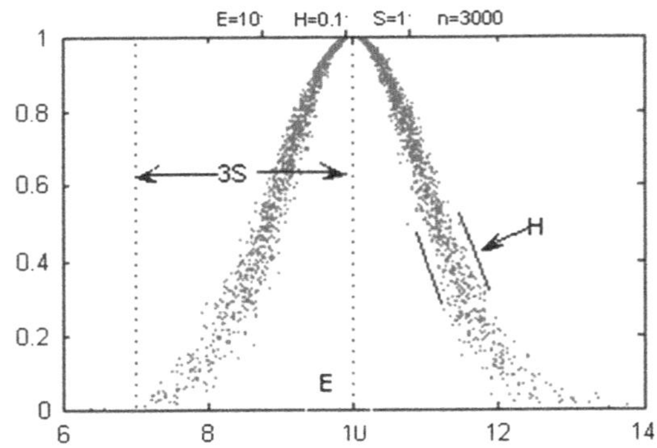

Fig. 4.3 Three digital characteristics of a normal cloud

proposed cloud adaptive particle swarm optimization (CAPSO); Zhang and Shao [17] execute variation for the particle with the cloud model, designed new complete cloud particle swarm optimization (NCCPSO) and cloud hyper-mutation particle swarm optimization (CHPSO), and eventually, all receive better performances.

Suppose T be the language value in the domain u, mapping $CT(x) : u \to [0, 1]$, $\forall x \in u, \quad x \to CT(x)$, then the distribution of $CT(x)$ on u is called the membership cloud under T. In case of obeying the normal distribution, $CT(x)$ is known as the normal cloud model [38]. The overall characteristics of the cloud model can be represented by the three digital features including desired E, entropy S, and hyper entropy H. The schematic for the normal cloud model is shown in Fig. 4.3.

The cloud theory is a model that contains the transferring procedure of uncertainty between quality concept and quantity data representation by using natural language [2]. It reflects the fuzziness and randomness of the concept within the affair and human knowledge in the objective world [18]. Let D be the language value of domain u and mapping $C_D(x) : u \to [0, 1], \quad \forall x \in u, \quad x \to C_D(x)$, then the distribution of $C_D(x)$ on u is called the membership cloud under D. If the distribution of $C_D(x)$ is normal, it is named a normal cloud model. It is a random number set with stable tendency that is distinguished by expectation E, entropy S, and super entropy H. They reflect the quantitative characteristics of concept $C_D(x)$ [2]. E is the expectation of spatial distribution of cloud droplet in the domain as well as the point that is the most able to represent the qualitative concept; S represents the measurable granularity of the qualitative concept, and the greater the entropy S is, usually the more macro the concept is. H is the uncertain measurement of entropy and is jointly determined by the randomness and fuzziness of entropy S. In case of knowledge representation, the cloud model has the characteristics of the uncertainty with certainty and the stability with change and reflects the basic principle of the evolution of species in nature. For the cloud model parameters, E represents the parent's good individual genetic characteristics and is the offspring's inheritance from the parent; entropy S and hyper entropy H indicate the uncertainty and fuzziness of inheritance process, showing the mutation characteristics of species in the evolutionary process. The algorithm or hardware for the generation of cloud droplets is called cloud droplet generator [2]. This book applies normal cloud generator to realize local search of the better individuals.

4.1.4 Normal Cloud Generator

As illustrated in Fig. 4.3, E determines the center of the cloud, S determines the range of the cloud, and, according to "3S" rule, about 99.74 % of the total cloud drops distribute between $[E - 3S, E + 3S]$; H determines the cloud drops dispersive degree, which means the larger the H is, the more dispersively the cloud drops locate [19].

A generator can generate a drop of cloud *drop* (x_i, u_0) with three digital characters E, S, H and a given certain u_0, which is called Y condition cloud generator, as shown as follows [20]:

Step 1: Input E, S, H, u_0, and n, the required number of drop.

Step 2: Generate a normal random number, S', by Eq. (4.6).

$$S' = \mathrm{randn}(S, H) \tag{4.6}$$

Step 3: Generate a new drop, x_i, by Eq. (4.7).

$$x_i = E \pm S' \sqrt{-2\ln(u_0)} \tag{4.7}$$

Step 4: Stop criteria. If the required number of drop is reached, otherwise, go back to step 2.

4.2 Chaotic Genetic Algorithm (CGA) in SVR's Parameters Determination

4.2.1 Shortcomings of GA and Improved by Chaotic Sequence

GA is auto-adaptive stochastic search techniques [21]; it generates new individuals with selection, crossover, and mutation operators. GA starts with a coding of the parameter set of all types of objective functions; thus, it has the ability to solve those traditional algorithms that are not easy to solve. GA has been employed in many empirical applications, due to their versatility and robustness in solving optimization problems. However, they have two major shortcomings when applied to GA, namely, slow convergence and becoming trapped in local optima, caused mainly by the reduced population diversity. The population diversity of an initial population cannot be maintained under selective pressure. Even if the initial individuals distributed uniformly, the qualities of the initial population are not necessarily uniformly arranged, meaning that the initial individuals are not necessarily fully diversified in the search space. Thus, most initial chromosomes are poor, and far from the global optimum. This is because if the initial population is not well designed, then the GA searches are always trapped into local optima.

For example, in Chap. 3, for SVRGA model, GA is able to reserve a few best fitted members of the whole population. However, the selection operation rules of GA mean only the few best fitting members of the whole population in a generation can survive. The population diversity is significantly reduced after some generations, meaning that GA might lead to a premature convergence to a local optimum in the searching for suitable parameters of an SVR model.

To overcome these drawbacks, some effective approaches and improvements on GA need to be discovered to maintain the population diversity and avoid leading to misleading local optimum. One possible approach is to divide the chromosome population into several subgroups and limit the crossover between the members in different subgroups in order to maintain the population diversity. However, this method requires a very large population size, which is not typical in business forecasting application problem solving. Another feasible scheme focuses on the chaos approach, due to its easy implementation and unique ability to avoid becoming trapped in local optima [2]. Chaos, defined as highly unstable motion in finite phase space, often occurs in deterministic nonlinear dynamic systems [3, 4]. Such motion is very similar to a random process ("randomicity"). Therefore, any variable in the chaotic space can travel ergodically over the whole space of interest ("ergodicity"). The variation of those chaotic variables has a delicate inherent rule, even though its variation looks like disorder ("regularity"). Additionally, it is extremely sensitive to the initial condition, which is an important property sometimes referred to as the so-called butterfly effect [5, 6]. Lorenz discovered that minute changes in initial conditions steered subsequent simulations toward radically different final stales when attempting to simulate a global weather system numerically. Based on the two advantages of the chaos, the chaotic optimization algorithm (COA) was proposed to solve complex function optimization [3]. The COA transforms the variable of problems from the solution space to chaos space and then performs a search to discover the solution by the three characteristics (*randomicity*, *ergodicity*, and *regularity*) of the chaotic variables. Yuan et al. [22] recently presented the chaotic genetic algorithm (CGA), which integrates GAs with chaotic mapping operator (CMO), to exploit the respective searching advantages of both models. First, the three characteristics of the chaotic variable are employed to distribute the individuals of subgenerations ergodically in the defined space and thus avoid the premature selection of individuals in later generations. Second, CGA also exploits the convergence characteristic of GA to overcome the randomness of the chaotic process and hence increase the probability of producing better optimized individuals and of finding the global optimal solution. A series application of CGA has also been proposed [23–25].

4.2.2 Operation Procedure of CGA

Conversely, traditional GA and related improvements have a common arrangement that completely ignore the individuals' experiences during their lifetime. This is because they are based on randomized searches, with no necessary connections between the current and next generations except for some controlling operators

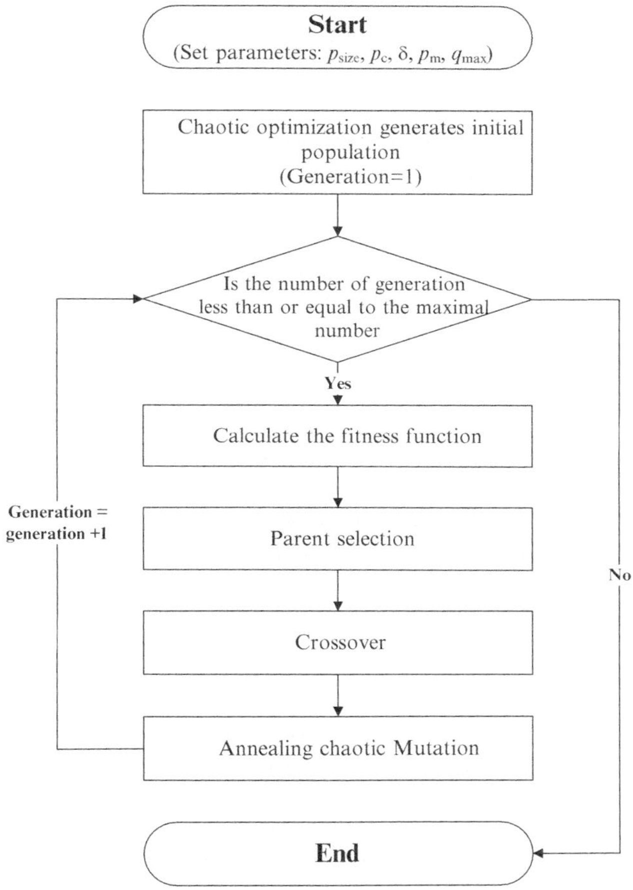

Fig. 4.4 The architecture of chaotic GA (CGA)

such as crossover and mutation operators. Mutation is an effective operator to increase and retain population diversity and is also an efficient approach to escape from local optima. Mutation can continuously pursue individuals of higher fitness value and guide the evolution of the whole population. However, the search is not exact, and the solution accuracy is poor, which means that a large scale of mutation is required to acquire the optimum solution in an extensive search. In contrast, if the precision is satisfactory, then the solution is often trapped in the local optima or takes too long to converge. Therefore, this book applies the annealing chaotic mutation operation, proposed by Cheng et al. [26]. The annealing chaotic mutation operation not only simulates the chaotic evolutionary process of biology but also employs chaotic variable to perform ergodic searching of the solution space, enabling it to find another better solution in the current neighborhood area of optimum solution, and to let GA possess ongoing motivity all along. Figure 4.4 shows the flowchart of the proposed procedure of CGA.

Step 1: Generate initial population by CMO. The values of the three parameters in an SVR model in the *i*th iteration can be represented as $X_k^{(i)}$, $k = \sigma$, C, ε. Set $i = 0$, and employ Eq. (4.8) to map the three parameters among the intervals (Min_k, Max_k) into chaotic variable $x_k^{(i)}$ located in the interval (0,1).

$$x_k^{(i)} = \frac{X_k^{(i)} - \text{Min}_k}{\text{Max}_k - \text{Min}_k}, \quad k = C, \ \sigma, \ \varepsilon \tag{4.8}$$

Then, adopt Eq. (4.5) with $\mu = 4$ to compute the next iteration chaotic variable, $x_k^{(i+1)}$. Transform $x_k^{(i+1)}$ to obtain three parameters for the next iteration, $X_k^{(i+1)}$, by the following Eq. (4.9):

$$X_k^{(i+1)} = \text{Min}_k + x_k^{(i+1)}(\text{Max}_k - \text{Min}_k) \tag{4.9}$$

After this transformation, the three parameters, σ, C, and ε, are encoded into a binary format and are represented by a chromosome composed of "genes" of binary numbers (Fig. 3.6). Each chromosome has three genes, respectively, representing three parameters. Each gene has 40 bits. For instance, if each gene contains 40 bits, then a chromosome contains 120 bits. More bits in a gene correspond to finer partitioning of the search space.

Step 2: Evaluate fitness. Evaluate the fitness (forecasting errors) of each chromosome. In this work, a negative mean absolute percentage error (MAPE) is used as the fitness function. The MAPE is calculated as Eq. (3.13).

Step 3: Selection. Based on fitness functions, chromosomes with higher fitness values are more likely to yield offspring in the next generation. The roulette wheel selection principle [21] is applied to choose chromosomes for reproduction.

Step 4: Crossover operations. In crossovers, chromosomes are paired randomly. The proposed scheme adopts the single-point-crossover principle. Segments of paired chromosomes between two determined breakpoints are swapped. For simplicity, suppose a gene has 4 bits, meaning that a chromosome contains 12 bits (Fig. 3.7). The values of the three parameters before crossover is performed are 1.5, 1.25, and 0.34375, respectively, in parent #1, and 0.625, 8.75, and 0.15625, respectively, in parent #2. The three values after crossover are 1.625, 3.75, and 0.40625, respectively, for offspring #1, and 0.5, 6.25, and 0.09375, respectively, for offspring #2. Finally, the three crossover parameters are decoded into a decimal format.

Step 5: Annealing chaotic mutation. For the *i*th iteration (generation) crossover population ($\hat{X}_k^{(i)}$, $k = \sigma, C, \varepsilon$) of current solution space (Min_k,Max_k) are mapped to chaotic variable interval [0,1] to form the crossover chaotic variable space $\hat{x}_k^{(i)}$, as Eq. (4.10),

$$\hat{x}_k^{(i)} = \frac{\hat{X}_k^{(i)} - \text{Min}_k}{\text{Max}_k - \text{Min}_k}, \quad k = C, \sigma, \varepsilon, i = 1, 2, \ldots, q_{\max}, \tag{4.10}$$

Table 4.1 Parameters determination of SVRCGA model

Nos. of fed-in data	Parameters σ	C	ε	MAPE of testing (%)
5	247.51	146.24	3.9383	3.726
10	6.37	116.48	1.1626	3.493
15	59.86	4,162.60	16.5860	3.444
20	357.68	7,827.20	4.1306	3.726
25	**22.18**	**6,705.30**	**21.8030**	**3.382**

where $q_{\max}$ is the maximum evolutional generation of the population. Then, the ith chaotic variable $x_k^{(i)}$ is summed up to $\hat{x}_k^{(i)}$, and the chaotic mutation variable is also mapped to interval [0,1] as in Eq. (4.11)

$$\tilde{x}_k^{(i)} = \hat{x}_k^{(i)} + \delta x_k^{(i)}, \tag{4.11}$$

where δ is the annealing operation. Finally, the chaotic mutation variable obtained in interval [0,1] is mapped to the solution interval (Min_k,Max_k) by definite probability of mutation (p_{m}), as shown in Eq. (4.12), thus completing a mutative operation.

$$\tilde{X}_k^{(i)} = \text{Min}_k + \tilde{x}_k^{(i)}(\text{Max}_k - \text{Min}_k) \tag{4.12}$$

Step 6: Stop condition. If the number of generations is equal to a given scale, then the best chromosomes are presented as a solution; otherwise, go back to step 2.

4.2.3 CGA for Three Parameters Determination and Forecasting Results

This subsection will demonstrate the three parameters determination of the proposed hybrid model (SVR with CGA), namely, SVRCGA model.

Similarly, in the training stage, the rolling-based forecasting procedure is also employed to obtain the forecasting load and receive training error in the training stage. Then, if training errors improvement occurs, the three kernel parameters—σ, C, and ε—of the SVRCGA model adjusted by CGA are employed to calculate the validation error. The adjusted parameters with minimum validation error are also selected as the most appropriate parameters. The forecasting results and the suitable parameters for the SVRSA model are illustrated in Table 4.1, in which it is also indicated that these two models all perform the best when 25 fed-in data are used.

Table 4.2 shows the actual values and the forecast values obtained using various forecasting models, ARIMA(1,1,1), GRNN(σ = 3.33), TF-ε-SVR-SA, SVRGA, and SVRCGA models. The MAPE values are calculated to compare fairly the proposed models with other alternative models. The proposed SVRCGA model has smaller MAPE values than ARIMA, GRNN, TF-ε-SVR-SA, and SVRGA models. Furthermore, to verify the significance of accuracy improvement of SVRCGA model

Table 4.2 Forecasting results of ARIMA, GRNN, TF-ε-SVR-SA, SVRGA, and SVRCGA models (unit: hundred million kWh)

Time point (month)	Actual	ARIMA(1,1,1)	GRNN($\sigma = 3.33$)	TF-ε-SVR-SA	SVRGA	SVRCGA
Oct. 2008	181.07	192.932	191.131	184.504	178.326	185.224
Nov. 2008	180.56	191.127	187.827	190.361	178.355	186.046
Dec. 2008	189.03	189.916	184.999	202.980	178.355	186.865
Jan. 2009	182.07	191.995	185.613	195.753	178.356	187.680
Feb. 2009	167.35	189.940	184.397	167.580	178.357	188.493
Mar. 2009	189.30	183.988	178.988	185.936	178.358	189.149
Apr. 2009	175.84	189.348	181.395	180.165	181.033	178.300
MAPE (%)		6.044	4.636	3.799	3.676	3.382

Table 4.3 Wilcoxon signed-rank test

	Wilcoxon signed-rank test	
	$\alpha = 0.025$	$\alpha = 0.05$
Compared models	$W = 2$	$W = 3$
SVRCGA vs. ARIMA(1,1,1)	1[a]	1[a]
SVRCGA vs. GRNN($\sigma = 3.33$)	3	3[a]
SVRCGA vs. TF-ε-SVR-SA	3	3[a]
SVRCGA vs. SVRGA	1[a]	1[a]

[a]Denote that SVRCGA model significantly outperforms other alternative models

Table 4.4 Asymptotic test

	Asymptotic (S_1) test	
Compared models	$\alpha = 0.05$	$\alpha = 0.10$
SVRCGA vs. ARIMA(1,1,1)	H_0: $e_1 = e_2$ $S_1 = -8.776$; $p = 0.000$ (reject H_0)	H_0: $e_1 = e_2$ $S_1 = -8.776$; $p = 0.000$ (reject H_0)
SVRCGA vs. GRNN($\sigma = 3.33$)	H_0: $e_1 = e_2$ $S_1 = -0.739$; $p = 0.2299$ (not reject H_0)	H_0: $e_1 = e_2$ $S_1 = -0.739$; $p = 0.2299$ (not reject H_0)
SVRCGA vs. TF-ε-SVR-SA	H_0: $e_1 = e_2$ $S_1 = 0.705$; $p = 0.2404$ (not reject H_0)	H_0: $e_1 = e_2$ $S_1 = 0.705$; $p = 0.2404$ (not reject H_0)
SVRCGA vs. SVRGA	H_0: $e_1 = e_2$ $S_1 = 16.415$; $p = 0.000$ (reject H_0)	H_0: $e_1 = e_2$ $S_1 = 16.415$; $p = 0.000$ (reject H_0)

comparing with ARIMA(1,1,1), GRNN($\sigma = 3.33$), TF-ε-SVR-SA, and SVRGA models, the Wilcoxon signed-rank test and Asymptotic test, as mentioned, are also conducted. The test results are shown in Tables 4.3 and 4.4, respectively. Clearly, the SVRCGA model is significantly superior to ARIMA and SVRGA models, except GRNN and TF-ε-SVR-SA models (only receives significance with $\alpha = 0.05$ level in Wilcoxon test, but all fails with both levels in asymptotic test). Figure 4.5 is provided to illustrate the forecasting accuracy among different models.

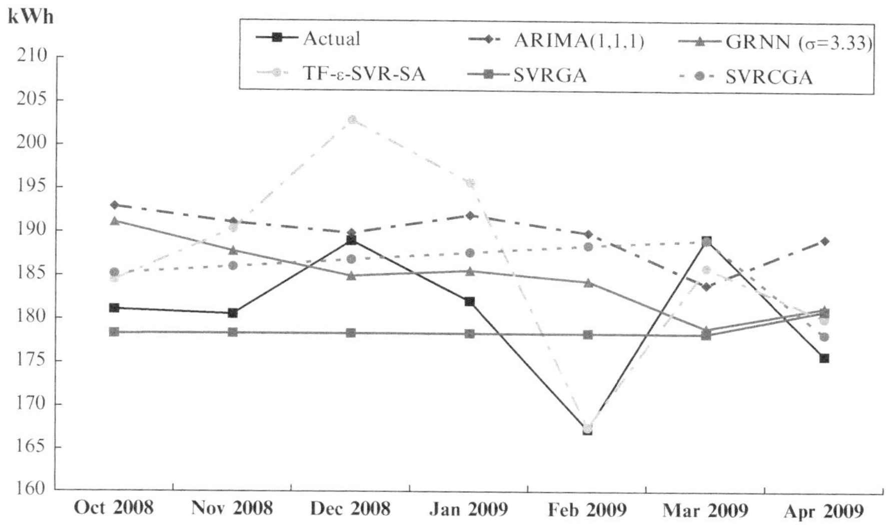

Fig. 4.5 Forecasting results of ARIMA, GRNN, TF-ε-SVR-SA, SVRGA, and SVRCGA models

The superior performance of the SVRCGA model is not only caused from several similar reasons of SVRGA model, such as SVR-based model with nonlinear mapping capabilities, minimizing the structural risks rather than the training errors, but also is caused from the chaotic mapping operator (CMO) of CGA and the searching capability of GA to determine the appropriate set of parameters. For example, based on Tables 4.1 and 3.4, the CGA is excellently to shift the local solution of SVRGA model $(\sigma, C, \varepsilon) = (686.16, 5048.40, 19.3170)$ with local optimal forecasting errors, in terms of MAPE (3.676), to another better solution $(\sigma, C, \varepsilon) = (22.18, 6705.30, 21.8030)$ of SVRCGA model to be the appropriate optimal forecasting error in terms of MAPE (3.382). Thus, it reveals that chaotic sequence is capable to avoid trapping in local minimum to improve forecasting accuracy. However, it is also clear that SVRCGA model is not fitting the actual electric loads very well but better than SVRGA model, even it has significant smaller MAPE values than other alternatives. Therefore, it also still requires hybridizing other novel mechanism to improve this shortcoming.

4.3 Chaotic Simulated Annealing Algorithm in SVR's Parameters Determination

4.3.1 Shortcomings of SA Algorithm and Improved by Chaotic Sequence

SA algorithm is a generic probabilistic search technique that simulates the material physical process of heating and controlled cooling. Each step of SA algorithm attempts to replace the current state by a random move. The new state may then be accepted with a probability that depends both on the difference between the corresponding function values and also on a global parameter, temperature. Thus, SA algorithm has some institution to reach more ideal solutions.

However, based on the operation procedure of SA algorithm, subtle and skillful adjustments in the annealing schedule are required, such as the size of the temperature steps during annealing, the temperature range, the number of restarts, and the redirection of the search [1]. In addition, due to its Monte Carlo scheme and lack of knowledge memory functions, time-consuming is also an another boring problem. To overcome these shortcomings, it is necessary to find some effective approach and improvement on SA's subtle and skillful adjustment in the annealing schedule. One feasible approach is focused on the chaos approach, due to its easy implementation and special ability to avoid being trapped in local optimum [1]. Chaos often occurs in a deterministic nonlinear dynamic system [3, 4]. Chen and Aihara [27] propose the concept of transiently chaotic dynamics, namely, chaotic simulated annealing (CSA) algorithm, which are temporarily generated for foraging and self-organizing, then, gradually vanished with autonomous decreasing of the temperature, and are accompanied by successive bifurcations and converged to a stable equilibrium. There are mainly two significant differences between CSA algorithm and SA algorithm. Firstly, SA is stochastic based on the Monte Carlo scheme, while CSA algorithm is deterministic with transiently chaotic dynamics. Secondly, the convergent processing of SA algorithm is undertaken by control of stochastic "thermal" fluctuations while that of CSA algorithm is by control of bifurcation structures. In addition, the searching region in CSA algorithm is usually very small due to its restricted dynamics to a fractal subspace with continuous state; thus, CSA algorithm can be expected to perform efficient searching including a global optimum state [27]. Henceforward, several applications of CSA algorithm have also been conducted [28, 29], but very sparse application in parameter determination. Therefore, CSA algorithm is employed to replace the stochastic "thermal" fluctuations control from traditional SA algorithm to optimize the parameter selection in an SVR model.

4.3.2 *Operation Procedure of CSA Algorithm*

To design the CSA algorithm, many principal factors, such as solution coding, initial temperature, and cooling schedule, are similar to the SA factors; more procedure details of the CSA algorithm on this study are as follows, and the flowchart is shown as Fig. 4.6.

Step 1: Initialization (to obtain the initial state). Let the values of the three parameters in an SVR model in the ith iteration be represented as $X_k^{(i)}$, $k = \sigma, C, \varepsilon$. Set $i = 0$ and employ the following equation [Eq. (4.13)] to map the three parameters among the intervals (Min_k,Max_k) into chaotic variables $x_k^{(i)}$ located in the interval (0,1):

$$x_k^{(i)} = \frac{X_k^{(i)} - \text{Min}_k}{\text{Max}_k - \text{Min}_k}, \quad k = \sigma, C, \varepsilon. \tag{4.13}$$

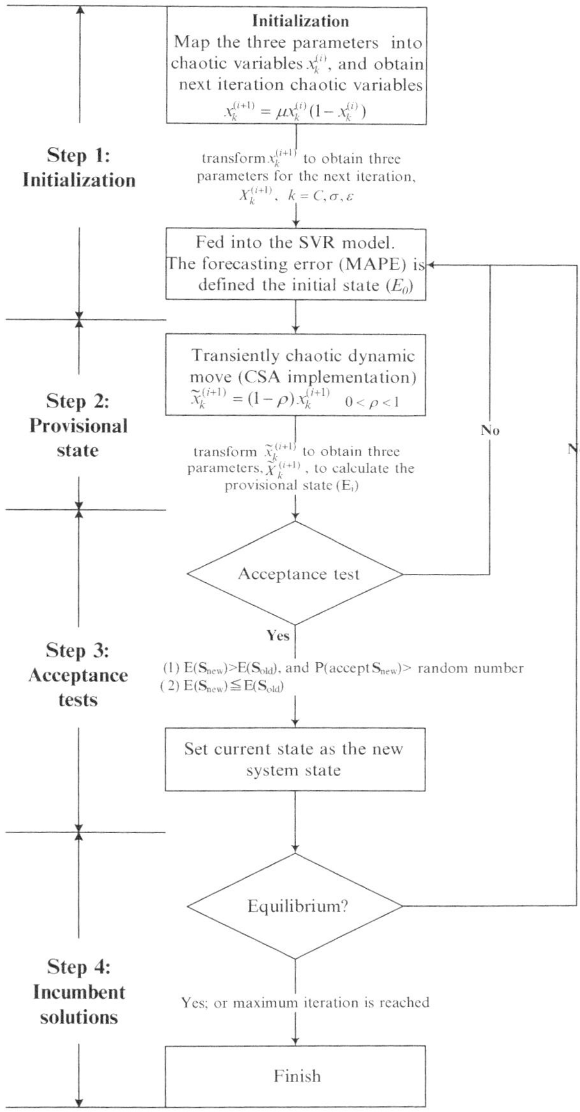

Fig. 4.6 The architecture of chaotic simulated annealing (CSA) algorithm

Then, employ the chaotic sequence, defined as Eq. (4.14), with $\mu = 4$ to compute the next iteration chaotic variables, $x_k^{(i+1)}$:

$$x_k^{(i+1)} = \mu x_k^{(i)}(1 - x_k^{(i)}), \tag{4.14}$$

where μ is the so-called bifurcation parameter of the system, $\mu \in [0,4]$. And, transform $x_k^{(i+1)}$ to obtain three parameters for the next iteration, $X_k^{(i+1)}$, by the following Eq. (4.15):

$$X_k^{(i+1)} = \text{Min}_k + x_k^{(i+1)}(\text{Max}_k - \text{Min}_k). \tag{4.15}$$

After this transformation, the three parameters, σ, C, and ε, are fed into the SVR model. The negative mean absolute percentage error (-MAPE) of forecasting error is defined as the system state (E_i), as given by Eq. (3.13). Finally, the initial state (E_0) is obtained by initial parameter combination.

Step 2: Provisional state. To obtain the provisional state, make a transiently chaotic dynamic move (i.e., CSA implementation) to change the existing system state (E_i) in the ith iteration to a provisional state. The three new chaotic variables (with simulated annealing effect) in the $i + 1$th iteration, $\tilde{x}_k^{(i+1)}$, are generated by using Eq. (4.16):

$$\tilde{x}_k^{(i+1)} = (1 - \rho)x_k^{(i+1)}, \tag{4.16}$$

where ρ $(0 < \rho < 1)$ governs the bifurcation speed of the transient chaos; for the larger value of ρ, $\tilde{x}_k^{(i+1)}$ decreases rapidly; therefore, the chaotic dynamics of $x_k^{(i+1)}$ vanishes quickly, on the contrary, for the smaller value of ρ, the chaotic dynamics of $x_k^{(i+1)}$ in turn lasts longer. In this investigation, ρ is set as 0.003 [27].

Then, transform $\tilde{x}_k^{(i+1)}$ to obtain three parameters for the next iteration, $\tilde{X}_k^{(i+1)}$, by the following Eq. (4.17):

$$\begin{aligned} \tilde{X}_k^{(i+1)} &= \text{Min}_k + \tilde{x}_k^{(i+1)}(\text{Max}_k - \text{Min}_k) \\ &= \text{Min}_k + (1 - \rho)x_k^{(i+1)}(\text{Max}_k - \text{Min}_k) \\ &= X_k^{(i+1)} - \rho \times (\text{Max}_k - \text{Min}_k) \times x_k^{(i+1)}. \end{aligned} \tag{4.17}$$

After this transformation, the three parameters, σ, C, and ε, are fed into the SVR model. The MAPE of forecasting error is defined as the provisional state.

Step 3: Acceptance tests. To determine the acceptance or rejection of provisional state (Metropolis et al. [30]) by Eq. (4.18),

$$\begin{cases} \text{Accept the provisional state, if } E(s_{\text{new}}) > E(s_{\text{old}}), \text{ and } p < P(\text{accept } s_{\text{new}}), 0 \le p \le 1. \\ \text{Accept the provisional state, if } E(s_{\text{new}}) \le E(s_{\text{old}}) \\ \text{Reject the provisional state, otherwise.} \end{cases} \tag{4.18}$$

In Eq. (4.18), the p is a random number to determine the acceptance of the provisional state; $P(\text{accept } s_{\text{new}})$ is shown as Eq. (3.15). If the provisional state is accepted, then set the provisional state as the current state.

Table 4.5 Parameters determination of SVRCSA model

Nos. of fed-in data	Parameters			MAPE of testing (%)
	σ	C	ε	
5	464.06	399.70	0.6891	4.289
10	43.75	40.36	1.8729	3.759
15	3.53	165.38	7.3935	3.941
20	3.02	1,336.70	9.8374	3.871
25	**74.08**	**1,751.00**	**2.6841**	**3.633**

Step 4: Incumbent solutions. If the provisional state is not accepted, then return to step 2. Furthermore, if the current state is not superior to the (initial) system state, then repeat steps 2 and 3 until the current state is superior to the system state and set the current state as new system state. Previous studies [31, 32] indicated that the maximum number of loops (N_{sa}) is 100d to avoid infinitely repeated loops, where d denotes the problem dimension. In this investigation, three parameters (σ, C, and ε) are used to determine the system states. Therefore, N_{sa} is state to 300.

4.3.3 CSA Algorithm for Three Parameters Determination and Forecasting Results

This subsection will demonstrate the three parameters determination of the proposed hybrid model (SVR with CSA), namely, SVRCSA model.

Similarly, in the training stage, the rolling-based forecasting procedure is also employed to obtain the forecasting load and receive training error in the training stage. Then, if training errors improvement occurs, the three kernel parameters—σ, C, and ε—of the SVRCSA model adjusted by CSA algorithm are employed to calculate the validation error. The adjusted parameters with minimum validation error are also selected as the most appropriate parameters. The forecasting results and the suitable parameters for the SVRCSA model are illustrated in Table 4.5, in which it is also indicated that these two models all perform the best when 25 fed-in data are used.

Table 4.6 shows the actual values and the forecast values obtained using various forecasting models, ARIMA(1,1,1), GRNN(σ = 3.33), TF-ε-SVR-SA, SVRSA, and SVRCSA models. The MAPE values are calculated to compare fairly the proposed models with other alternative models. The proposed SVRCSA model has smaller MAPE values than ARIMA, GRNN, TF-ε-SVR-SA, and SVRSA models. Furthermore, to verify the significance of accuracy improvement of SVRCSA model comparing with ARIMA(1,1,1), GRNN(σ = 3.33), TF-ε-SVR-SA, and SVRSA models, the Wilcoxon signed-rank test and asymptotic test, as mentioned, are also conducted. The test results are shown in Tables 4.7 and 4.8, respectively. Clearly, the SVRCSA model is significantly superior to ARIMA, TF-ε-SVR-SA, and SVRSA models and minor significantly superior to GRNN model

Table 4.6 Forecasting results of ARIMA, GRNN, TF-ε-SVR-SA, SVRSA, and SVRCSA models (unit: hundred million kWh)

Time point (month)	Actual	ARIMA(1,1,1)	GRNN($\sigma = 3.33$)	TF-ε-SVR-SA	SVRSA	SVRCSA
Oct. 2008	181.07	192.932	191.131	184.504	184.584	184.059
Nov. 2008	180.56	191.127	187.827	190.361	185.412	183.717
Dec. 2008	189.03	189.916	184.999	202.980	185.557	183.854
Jan. 2009	182.07	191.995	185.613	195.753	185.593	184.345
Feb. 2009	167.35	189.940	184.397	167.580	185.737	184.489
Mar. 2009	189.30	183.988	178.988	185.936	184.835	184.186
Apr. 2009	175.84	189.348	181.395	180.165	184.390	184.805
MAPE (%)		6.044	4.636	3.799	3.801	3.633

Table 4.7 Wilcoxon signed-rank test

	Wilcoxon signed-rank test	
	$\alpha = 0.025$	$\alpha = 0.05$
Compared models	$W = 2$	$W = 3$
SVRCSA vs. ARIMA(1,1,1)	1[a]	1[a]
SVRCSA vs. GRNN($\sigma = 3.33$)	3	3[a]
SVRCSA vs. TF-ε-SVR-SA	2[a]	2[a]
SVRCSA vs. SVRSA	1[a]	1[a]

[a]Denote that SVRCSA model significantly outperforms other alternative models

Table 4.8 Asymptotic test

	Asymptotic (S_1) test	
Compared models	$\alpha = 0.05$	$\alpha = 0.10$
SVRCSA vs. ARIMA(1,1,1)	H_0: $e_1 = e_2$ $S_1 = -9.659$; $p = 0.000$ (reject H_0)	H_0: $e_1 = e_2$ $S_1 = -5.727$; $p = 0.000$ (reject H_0)
SVRCSA vs. GRNN($\sigma = 3.33$)	H_0: $e_1 = e_2$ $S_1 = -1.544$; $p = 0.0613$ (not reject H_0)	H_0: $e_1 = e_2$ $S_1 = -1.544$; $p = 0.0613$ (not reject H_0)
SVRCSA vs. TF-ε-SVR-SA	H_0: $e_1 = e_2$ $S_1 = -1.824$; $p = 0.0341$ (reject H_0)	H_0: $e_1 = e_2$ $S_1 = -1.824$; $p = 0.0341$ (reject H_0)
SVRCSA vs. SVRSA	H_0: $e_1 = e_2$ $S_1 = -6.106$; $p = 0.000$ (reject H_0)	H_0: $e_1 = e_2$ $S_1 = -6.106$; $p = 0.000$ (reject H_0)

(only receives significance with $\alpha = 0.05$ level in Wilcoxon test, and with $\alpha = 0.10$ level in asymptotic test). Figure 4.7 is provided to illustrate the forecasting accuracy among different models.

The superior performance of the SVRCSA model is not only caused from several similar reasons of SVRSA model, such as SVR-based model with nonlinear mapping capabilities, minimizing the structural risks rather than the training errors the

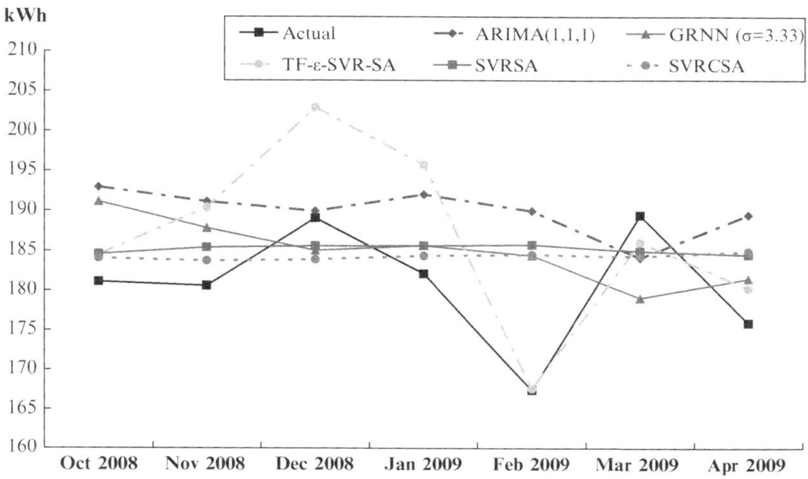

Fig. 4.7 Forecasting results of ARIMA, GRNN, TF-ε-SVR-SA, SVRSA, and SVRCSA models

quadratic programming technique which is based on the assumptions of convex set and existence of global optimum solution, but also is caused from the chaotic mapping operator (CMO) of CSA algorithm and the proper determination of SVR parameters by SA algorithm. Thus, it should be theoretically approximated to the global optimum solution if the superior searching algorithm is employed. For example, based on Tables 4.5 and 3.8, the CSA algorithm is excellently to shift the local solution of SVRSA model, $(\sigma, C, \varepsilon) = (94.998, 9435.20, 12.6570)$ with local optimal forecasting errors, in terms of MAPE (3.801), to another better solution, $(\sigma, C, \varepsilon) = (74.08, 1751.00, 2.6841)$ of SVRCSA model to be the appropriate optimal forecasting error in terms of MAPE (3.633). Thus, it reveals that chaotic sequence is capable to avoid trapping in local minimum to achieve forecasting accuracy improvement by integrated into the SVR model. However, it is also clear that SVRCSA model is not fitting the actual electric loads very well but better than SVRSA model, even it has significant smaller MAPE values than other alternatives. Therefore, it also still requires hybridizing other novel mechanism to improve this shortcoming.

4.4 Chaotic Cloud Simulated Annealing Algorithm in SVR's Parameters Determination

4.4.1 Shortcomings of CSA Algorithm and Improved by Cloud Theory

As presented in Sects. 3.2 and 3.3, chaotic sequence is employed to transform these hyper-parameters to chaotic space to realize searching ergodically over the whole space. However, most chaotic sequences adopt logistic mapping function which is distributed at both ends in the interval [0,1]; it could not excellently strengthen the chaotic distribution characteristics. Comparing the analysis on chaotic distributed characteristics (mapping hyper-parameters into chaotic space), the cat mapping function is with good ergodic uniformity in the interval [0,1] and is not easily to fall into minor cycle [33] (see Fig. 4.1). This section decides to apply the cat mapping function to map hyper-parameters into chaotic space.

On the other hand, the chaotic sequence is also only employed to transform the three parameters of an SVR model from the solution space to chaotic space; any parameter in this kind of chaotic space can travel ergodically over the whole space of interest, eventually, to find out improved solution. Several disadvantages embedded in SA algorithm are required to be improved to get more accurate forecasting performance of an SVR model. For example, as mentioned that subtle and skillful adjustment in the annealing schedule is required, particularly, the temperature of each state is discrete and unchangeable, which does not meet the requirement of continuous decrease in temperature in actual physical annealing processes. In addition, SA algorithm is easy to accept deteriorate solution with high temperature, and it is hard to escape from local minimum trap with low temperature [19]. Finally, due to its Monte Carlo scheme and lack of knowledge memory functions, time-consuming is also another boring problem of SA algorithm.

Cloud theory is a model of the uncertainty transformation between quantitative representation and qualitative concept using language value [2]. It is successfully used in intelligence control [13, 34], data mining [35], spatial analysis [36], intelligent algorithm improvement [37], and so on. Meanwhile, several disadvantages from evolutionary algorithms themselves should also be considered any feasible improvements to receive higher forecasting accuracy level. In case of SA algorithm, its basic operation procedure requires subtle and skillful adjustment during annealing schedule such as the temperature decreasing size. Especially, most applications disregard that temperature should be decreased continuously rather discretely and fixed for each state in actual physical annealing processes. Moreover, based on the theoretical definition of SA algorithm, in the high-temperature stage, it is easy to accept worsened solution than to converge to local minimum while decreasing to low temperature [19]. Based on SA's mechanism, along with the temperature decreasing, the annealing process will, like a fuzzy system, let these hyper-parameters move from large scale to small scale randomly as the temperature decreases. Cloud theory can transform the uncertainty between quantitative representation and qualitative concept (in language form) [2], that is, it can successfully realize the transformation between a qualitative concept in words and numerical representation [2]. Therefore, it is suitable to be employed to solve the problem of temperature decreasing discretely. This investigation tries to apply the chaotic sequence (by cat mapping function) with cloud theory-based simulated annealing algorithm (namely CCSA) to determine the values of these hyper-parameters in an SVR model to escape from stagnation.

4.4.2 Operation Procedure of CCSA Algorithm

In previous section (Sect. 4.3), author has tried to employ CSA algorithm to overcome these shortcomings, in which, the transiently chaotic dynamics are temporarily generated for foraging and self-organizing, then gradually vanished with autonomous decreasing of the temperature, and are accompanied by successive bifurcations and converged to a stable equilibrium. Therefore, CSA algorithm has significantly improved the randomization of Monte Carlo scheme, has controlled the convergent process by bifurcation structures instead of stochastic "thermal" fluctuations, and, eventually, performed efficient searching including a global optimum state [27]. However, as mentioned that the temperature of each state is discrete and unchangeable, which does not meet the requirement of temperature continuously decrease in actual physical annealing processes. Even some temperature annealing function is exponential in general [38], the temperature is gradually fallen with a fixed value in every annealing step, and the changing process of temperature between two neighbor steps is not continuous. This phenomenon also appears while other types of temperature update functions are implemented, such as arithmetical, geometrical, or logarithmic one [39]. In the cloud theory, by introducing the *Y* condition normal cloud generator to the temperature generation process, it can randomly generate a group of new values that distribute around the given value like "cloud." Let the fixed temperature point of each step become a changeable temperature zone, the temperature of each state generation in every annealing step is chosen randomly, the course of temperature changing in the whole annealing process is nearly continuous and fits the physical annealing process better [19]. Therefore, based on chaotic sequence and cloud theory, the CCSA algorithm is employed to replace the stochastic "thermal" fluctuations control from traditional SA algorithm, to enhance the continuously physical temperature annealing process from CSA algorithm, and to optimize the parameter selection in an SVR model.

For strengthening chaotic distribution characteristics, most chaotic sequences adopt logistic mapping function as the chaotic sequence generator. However, the logistic mapping function is mostly distributed at both ends due to complying with Chebyshev distribution of more at both ends and less in middle. The cat mapping function is relatively uniform in the interval [0,1] and has no cyclic phenomenon during its iteration process. This section tries to employ the cat mapping function, with good ergodic uniformity and not easily to fall into minor cycle, to the CSA algorithm.

To design the CCSA algorithm, many principal factors, such as solution coding and initial temperature, but annealing process, are similar to the SA factors, more procedure details of the CCSA algorithms on this study are as follows, and the flowchart is shown as Fig. 4.8.

Step 1: Initialization. To unify the initial temperature and to obtain the initial state.

The initial temperature (T_0) has to be unified, initially we spot a value of 0.1 for the temperature (i.e., $T_0 = 0.1$). The initial state would be obtained after mapping three parameters into chaotic variables. Let the values of the three parameters in an

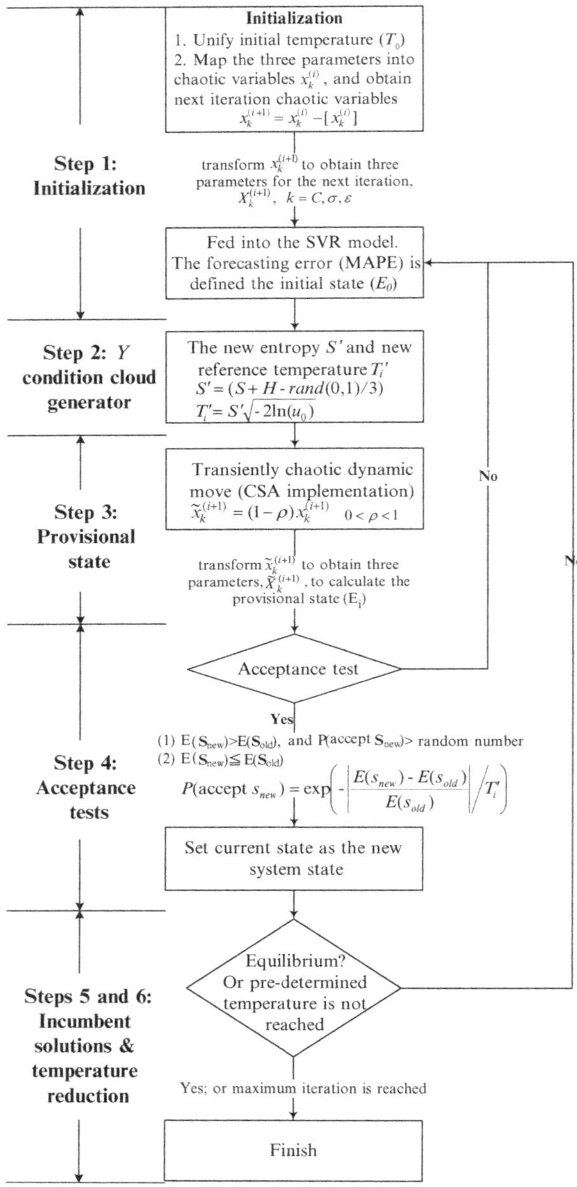

Fig. 4.8 The architecture of chaotic cloud simulated annealing (CCSA) algorithm

SVR model for the ith iteration be represented as $X_k^{(i)}$, $k = \sigma, C, \varepsilon$. Set $i = 0$ and employ the following equation [Eq. (4.13)] to map the three parameters among the intervals (Min_k,Max_k) into chaotic variables $x_k^{(i)}$ located in the interval (0,1).

Then, employ the chaotic sequence, defined as Eq. (4.19), to compute the next iteration chaotic variables:

$$x_k^{(i+1)} = x_k^{(i)} - \left[x_k^{(i)}\right]. \tag{4.19}$$

And, transform $x_k^{(i+1)}$ to obtain three parameters for the next iteration, $X_k^{(i+1)}$, by the following Eq. (4.15).

After this transformation, the three parameters, σ, C, and ε, are fed into the SVR model. The mean absolute percentage error (MAPE) of forecasting error is defined as the system state (E_i), as given by Eq. (3.13). Finally, the initial state (E_0) is obtained by initial parameter combination.

Step 2: Y condition cloud generator. Set initial values of the characters S, H, and u_0 in the Y condition cloud generator. Let $S = T_i$, $H = T_i$ and $u_0 = 1.0 - T_i$; then, new entropy S' and new reference temperature T_i' will be calculated by Eqs. (4.20) and (4.21), respectively,

$$S' = (S + H - \text{rand}(0, 1)/3), \tag{4.20}$$

$$T'_i = S'\sqrt{-2\ln(u_0)}. \tag{4.21}$$

Entropy S and super entropy H are set as the same reference temperature T_i, due to their positively correlations with the temperature stage in cloud theory. Since S determines the range of the cloud and H determines the cloud drops' dispersive degree, therefore, at the high temperature stage, if the changeable range of annealing temperature is much wider and drop is more dispersive, sufficient randomness of the annealing process will be ensured. On the contrary, at the low temperature stage, if the changeable range of annealing temperature is much narrower and less dispersive, the stable tendency of annealing process will be guaranteed.

Step 3: Provisional state. To obtain the provisional state, make a transiently chaotic dynamic move (i.e., CSA implementation) to change the existing system state (E_i) in the ith iteration to a provisional state. The three new chaotic variables (with simulated annealing effect) in the $i + 1$th iteration, $\tilde{x}_k^{(i+1)}$, are generated by using Eq. (4.16). Then, transform $\tilde{x}_k^{(i+1)}$ to obtain three parameters for the next iteration, $\tilde{X}_k^{(i+1)}$, by the following Eq. (4.17). After this transformation, the three parameters, σ, C, and ε, are fed into the SVR model. The MAPE of forecasting error is defined as the provisional state.

Step 4: Acceptance tests. To determine the acceptance or rejection of provisional state (Metropolis et al. [30]) by Eq. (4.18). For CCSA algorithm, $P(\text{accept } s_{\text{new}})$, the probability of accepting the new state is given by the following probability function, Eq. (4.22), to determine the acceptance of the provisional state:

$$P(\text{accept } s_{\text{new}}) = \exp\left(-\left|\frac{E(s_{\text{new}}) - E(s_{\text{old}})}{E(s_{\text{old}})}\right| / T'_i\right). \tag{4.22}$$

If the provisional state is accepted, then set the provisional state as the current state.

Step 5: Incumbent solutions. If the provisional state is not accepted, then return to step 2. Furthermore, if the current state is not superior to the (initial) system state, then repeat steps 2–4 for next iteration (i.e., $i = i + 1$) until the current state is superior to the system state, and set the current state as new system state. Previous studies [31, 32] indicated that the maximum number of loops (N_{sa}) is $100d$ to avoid infinitely repeated loops, where d denotes the problem dimension. In this investigation, three parameters (σ, C, ε) are used to determine the system states. Therefore, N_{sa} is state to 300.

Step 6: Temperature reduction. After the new system state is obtained, reduce the temperature. The temperature update function in CCSA algorithm is exponential in general. Let the annealing index (ρ) be the same in step 3, and the step counter is i, then the new temperature in step i is computed according to Eq. (4.23):

$$T = T_0 \times \rho^i. \tag{4.23}$$

If the predetermined temperature is reached, then stop the algorithm and the latest state is an approximate optimal solution. Otherwise, go to step 2.

Based on the temperature update function, the annealing temperature is gradually falling, that is, at the end of each step, it changes according to Eq. (4.23) as base temperature for next step. Then, by employing the Y condition cloud generator and taking base temperature as a certain value, the new randomly generated values would distribute around the base temperature and look like a "cloud." It means that during this temperature generation process, the fixed temperature point of each step becomes a changeable temperature zone; the course of temperature changing is nearly continuous and fits the physical annealing process better.

4.4.3 CCSA Algorithm for Three Parameters Determination and Forecasting Results

This subsection will demonstrate the three parameters determination of the proposed hybrid model (SVR with CCSA), namely, SVRCCSA model.

Similarly, in the training stage, the rolling-based forecasting procedure is also employed to obtain the forecasting load and receive training error in the training stage. Then, if training errors improvement occurs, the three kernel parameters—σ, C, and ε—of the SVRCCSA model adjusted by CCSA algorithm are employed to calculate the validation error. The adjusted parameters with minimum validation error are also selected as the most appropriate parameters. The forecasting results and the suitable parameters for the SVRCCSA model are illustrated in Table 4.9, in which it is also indicated that these two models all perform the best when 25 fed-in data are used.

Table 4.10 shows the actual values and the forecast values obtained using various forecasting models: ARIMA(1,1,1), GRNN($\sigma = 3.33$), TF-ε-SVR-SA, SVRSA, SVRCSA, and SVRCCSA models. The MAPE values are calculated to

Table 4.9 Parameters determination of SVRCCSA model

Nos. of fed-in data	Parameters			MAPE of testing (%)
	σ	C	ε	
5	903.83	357.61	2.0357	4.475
10	4.03	132.04	1.4276	4.040
15	196.56	47.86	6.1689	3.828
20	248.61	9467.10	13.3460	3.574
25	**930.43**	**1,737.40**	**17.8360**	**3.406**

Table 4.10 Forecasting results of ARIMA, GRNN, TF-ε-SVR-SA, SVRSA, SVRCSA, and SVRCCSA models (unit: hundred million kWh)

Time point (month)	Actual	ARIMA (1,1,1)	GRNN ($\sigma = 3.33$)	TF-ε-SVR-SA	SVRSA	SVRCSA	SVRCCSA
Oct. 2008	181.07	192.932	191.131	184.504	184.584	184.059	179.138
Nov. 2008	180.56	191.127	187.827	190.361	185.412	183.717	179.789
Dec. 2008	189.03	189.916	184.999	202.980	185.557	183.854	179.834
Jan. 2009	182.07	191.995	185.613	195.753	185.593	184.345	179.835
Feb. 2009	167.35	189.940	184.397	167.580	185.737	184.489	179.835
Mar. 2009	189.30	183.988	178.988	185.936	184.835	184.186	179.835
Apr. 2009	175.84	189.348	181.395	180.165	184.390	184.805	182.514
MAPE (%)		6.044	4.636	3.799	3.801	3.633	3.406

Table 4.11 Wilcoxon signed-rank test

Compared models	Wilcoxon signed-rank test	
	$\alpha = 0.025$	$\alpha = 0.05$
	W = 2	W = 3
SVRCCSA vs. ARIMA(1,1,1)	0[a]	0[a]
SVRCCSA vs. GRNN($\sigma = 3.33$)	2[a]	2[a]
SVRCCSA vs. TF-ε-SVR-SA	2[a]	2[a]
SVRCCSA vs. SVRSA	0[a]	0[a]
SVRCCSA vs. SVRCSA	0[a]	0[a]

[a]Denote that SVRCCSA model significantly outperforms other alternative models

compare fairly the proposed models with other alternative models. The proposed SVRCCSA model has smaller MAPE values than ARIMA, GRNN, TF-ε-SVR-SA, SVRSA, and SVRCSA models. Furthermore, to verify the significance of accuracy improvement of SVRCCSA model comparing with ARIMA(1,1,1), GRNN ($\sigma = 3.33$), TF-ε-SVR-SA, SVRSA, and SVRCSA models, the Wilcoxon signed-rank test and asymptotic test, as mentioned, are also conducted. The test results are shown in Tables 4.11 and 4.12, respectively. Clearly, the SVRCCSA model is significantly superior to ARIMA(1,1,1), GRNN($\sigma = 3.33$), TF-ε-SVR-SA, SVRSA, and SVRCSA models. Figure 4.9 is provided to illustrate the forecasting accuracy among different models.

Table 4.12 Asymptotic test

Compared models	Asymptotic (S_1) test $\alpha = 0.05$	$\alpha = 0.10$
SVRCCSA vs. ARIMA(1,1,1)	H_0: $e_1 = e_2$ $S_1 = -11.723$; $p = 0.000$ (reject H_0)	H_0: $e_1 = e_2$ $S_1 = -11.723$; $p = 0.000$ (reject H_0)
SVRCCSA vs. GRNN($\sigma = 3.33$)	H_0: $e_1 = e_2$ $S_1 = -2.321$; $p = 0.0101$ (reject H_0)	H_0: $e_1 = e_2$ $S_1 = -2.321$; $p = 0.0101$ (reject H_0)
SVRCCSA vs. TF-ε-SVR-SA	H_0: $e_1 = e_2$ $S_1 = -2.873$; $p = 0.0020$ (reject H_0)	H_0: $e_1 = e_2$ $S_1 = -2.873$; $p = 0.0020$ (reject H_0)
SVRCCSA vs. SVRSA	H_0: $e_1 = e_2$ $S_1 = -12.077$; $p = 0.000$ (reject H_0)	H_0: $e_1 = e_2$ $S_1 = -12.077$; $p = 0.000$ (reject H_0)
SVRCCSA vs. SVRCSA	H_0: $e_1 = e_2$ $S_1 = -12.258$; $p = 0.000$ (reject H_0)	H_0: $e_1 = e_2$ $S_1 = -12.258$; $p = 0.000$ (reject H_0)

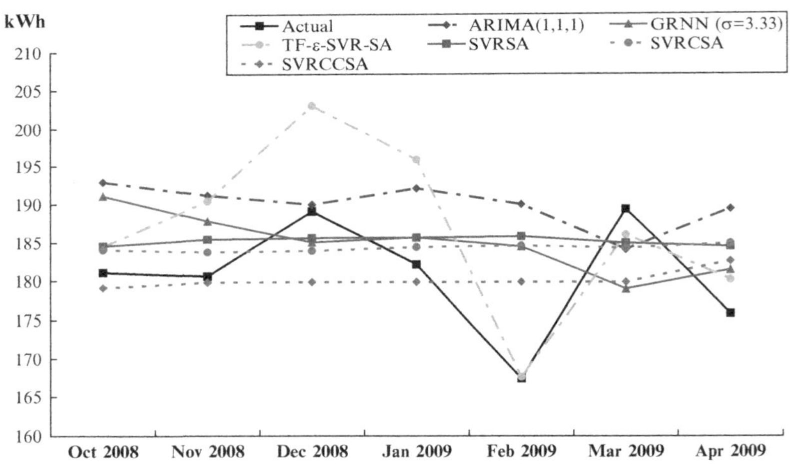

Fig. 4.9 Forecasting results of ARIMA, GRNN, TF-ε-SVR-SA, SVRSA, SVRCSA, and SVRCCSA models

The superior performance of the SVRCCSA model is not only caused from several reasons causes of SVRSA model, such as SVR-based model with nonlinear mapping capabilities, minimizing the structural risks rather than the training errors and the quadratic programming technique which is based on the assumptions of convex set and existence of global optimum solution, but also is caused from the chaotic mapping operator (CMO) of CSA algorithm, the proper determination of SVR parameters by SA algorithm, and the Y condition normal cloud generator to the temperature generation during the annealing process. Particularly for the

Y condition normal cloud generator, it can randomly generate a group of new values that distribute around the given value like "cloud" to obviously ensure temperature reducing continuously and to overcome the dilemma of the original SA algorithm; it is easy to accept worsened solution than to converge to local minimum while decreasing to low temperature. Therefore, it will be theoretically approximated to the global optimum solution if the superior searching algorithm is used. For example, based on Tables 4.5 and 4.9, the CSA algorithm is excellently to shift the local solution of SVRCSA model, $(\sigma, C, \varepsilon) = (74.08, 1751.00, 2.6841)$ with local optimal forecasting errors, in terms of MAPE (3.633), to another better solution, $(\sigma, C, \varepsilon) = (930.43, 1737.40, 17.8360)$ of SVRCCSA model to be the appropriate optimal forecasting error in terms of MAPE (3.406). Thus, it reveals that chaotic sequence is capable to avoid trapping in local minimum to achieve forecasting accuracy improvement by integrating into the SVR model. However, it is also clear that SVRCCSA model is also not fitting the actual electric loads very well but better than SVRSA and SVRCSA models, even if it has significant smaller MAPE values than other alternatives. Therefore, it also still requires hybridizing other novel mechanism to improve this shortcoming.

4.5 Chaotic GASA (CGASA) Algorithm in SVR's Parameters Determination

4.5.1 Shortcomings of GASA Algorithm and Improved by Chaotic Sequence

As indicated in Chap. 3, Sect. 3.5, the common drawback of GA and SA algorithm is premature convergence. Hybridization of genetic algorithm with simulated annealing (GASA) algorithm is an innovative trial by applying the superior capability of SA algorithm to reach more ideal solutions and by employing the mutation process of GA to enhance searching process. To continue improving to receive more accurate forecasting performance, it is deserved to find some novel mechanism to overcome these shortcomings from GA and SA algorithm. This section will also apply chaotic sequence to carefully expand variable searching space, i.e., let variable travel ergodically over the searching space. Therefore, the proposed chaotic genetic algorithm-simulated annealing (CGASA) algorithm, namely, SVRCGASA, is proposed to be expected to receive more accurate performance.

4.5.2 Operation Procedure of CCSA Algorithm

The proposed CGASA algorithm includes two parts: the CGA part and the SA algorithm part. The brief flow of CGASA algorithm is as follows: firstly, the CGA will evaluate the initialized population and implement three basic operators to

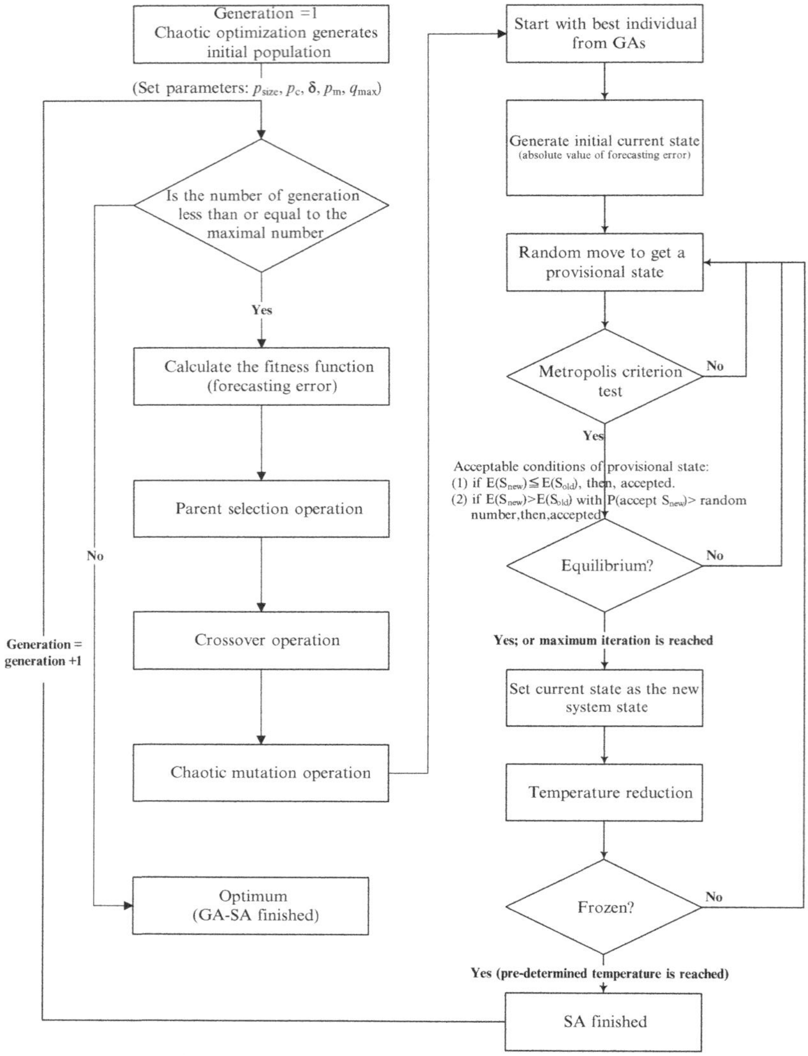

Fig. 4.10 The architecture of chaotic genetic-simulated annealing (CGASA) algorithm

generate new population and find out the best individual; secondly, deliver the determined best individual to SA algorithm for further processing; thirdly, while all SA processes are finished, the modified individual will be sent back to CGA for the next generation; fourthly, the computing iteration will be never stopped till the stop criteria of the algorithm are reached. The proposed procedure of CGASA algorithm is illustrated as follows and the flowchart is shown as Fig. 4.10.

4.5.2.1 The Procedure of the CGA Part

The worst shortcoming in the processes of GA is that the population diversity will reduce after several generation calculations, and then, it will lead to premature convergence (be trapped in local optima). The reduced population diversity is mainly caused of initial population not fully diversified in the search space and far from the global optimum even if the initial individuals distributed uniformly. Therefore, in this chapter, the chaotic mapping operator (CMO) will replace the random approach to generate the initial population.

Secondly, another shortcoming of GA is its ineffective operator, crossover operator, or mutation operator, which will lead to time-consuming mutation operator and slow convergent speed. Therefore, the suitable modifications for these two operators will highly help to improve the performance of GA. In this paper, the chaotic mutation operation, proposed by Cheng et al. [26], is employed, which uses chaotic variable to find another better solution in the current neighborhood area of optimum solution, and lets GA possess ongoing motivity all along. The procedure of CGA part is illustrated as follows:

Step 1: Initial population. Set three hyper-parameters of an SVR model as $X_k^{(i)}$, $k = \sigma, C$, and ε; i denotes the iteration number. Let $i = 0$, normalize the parameters as chaotic variables $x_k^{(i)}$ within the interval [0,1] by Eq. (4.8). Then, calculate and code the original chaotic variables $x_k^{(i)}$ in decimal format for the next iteration, $X_k^{(i+1)}$, by Eq. (4.5) with $\mu = 4$ and Eq. (4.9). The obtained (in decimal format) parameters, $X_k^{(i+1)}$, can be coded into a binary format and be presented as a chromosome composed of three genes; each gene has 40 bits.

Step 2: Fitness evaluation. Evaluate the fitness of the objective function with regard to each chromosome. In this chapter, the negative mean absolute percentage error (-MAPE), that is, the forecasting error of the trained SVR model, is used to evaluate the fitness. The MAPE is computed as Eq. (3.13).

Step 3: Selection operation. Chromosomes with small MAPE values are potential to generate offspring in the next generation. The roulette wheel selection principle [21] is employed while chromosomes are selected to be reproduced.

Step 4: Crossover operation. While crossover operating, chromosomes are randomly selected to be paired with each other. By using the single-point-crossover principle, segments of the paired chromosomes after the decided breakpoints are exchanged. Eventually, three parameters after crossover operator should be decoded back to the decimal format for next step.

Step 5: Chaotic mutation operation. Normalize the after-crossover parameters, $\hat{X}_k^{(i)}$, as chaotic variables $\hat{x}_k^{(i)}$, $k = C, \sigma, \varepsilon$, by Eq. (4.10), in which $q_{\max}$ is the maximum generation of the population and it is set to 500 in this section. Then, the ith chaotic mutation variables, $\hat{x}_k^{(i)}$, are calculated by Eq. (4.11), in which δ is the annealing mutation operation and is set to 0.9 in this section [40]. Finally, the obtained chaotic mutation variables will be transformed back with definite probability of mutation (p_{m}, will be set as 0.1 in this section) as Eq. (4.12).

Step 6: Stop criterion. If the number of generations is reached to the stop criterion, then the best individual could be obtained and delivered to the SA part; otherwise, go back to step 2.

4.5.2.2 The Procedure of the SA Part

In the proposed CGASA algorithm process, SA algorithm will conduct further processes on the best individual delivered from CGA. After the best individual being further improved, SA algorithm will send it back to CGA for the next generation. These computing iterations will never be stopped till the termination condition of the proposed algorithm is reached. The procedure of the SA algorithm part is illustrated as follows:

Step 1: Initial state. Calculate MAPE by the three received parameters from CGA, as shown by Eq. (3.13). The calculated MAPE is defined as the system state (E). Here, the initial state (E_0) is obtained.

Step 2: Provisional state. Adjust the existing system state to a provisional state by making a random move. Another set of three parameters are generated in this step.

Step 3: Metropolis criterion test. The following Metropolis criterion equation [Eq. (4.18)] is employed to determine the acceptance or rejection of provisional state [30]. If the provisional state is accepted, then set the provisional state as the current state.

Step 4: Incumbent solutions. If the provisional state is not accepted, then return to step 2. Furthermore, if the current state is not superior to the system state, then repeat steps 2 and 3 until the current state is superior to the system state and set the current state as new system state. Previous studies [31, 32] have recommended that the maximum number of loops (N_{sa}) is $100d$ to avoid infinitely repeated loops, where d denotes the problem dimension. In this study, three parameters are employed to determine the system states. Therefore, N_{sa} is set as 300.

Step 5: Temperature reduction. After the new system state is obtained, reduce the temperature. The new temperature reduction is obtained by Eq. (3.17), in which the ρ is recommended to be set as 0.9 in this study [40]. If the predetermined temperature is reached, then stop this SA part and the latest state is the approximate optimal solution; otherwise, go to step 2.

4.5.3 CGASA Algorithm for Three Parameters Determination and Forecasting Results

This subsection will demonstrate the three parameters determination of the proposed hybrid model (SVR with CGASA), namely, SVRCGASA model.

Table 4.13 Parameters determination of SVRCGASA model

Nos. of fed-in data	Parameters σ	C	ε	MAPE of testing (%)
5	88.16	457.74	3.7395	4.936
10	6.877	112.58	3.7047	3.834
15	6.379	142.85	0.0320	4.197
20	247.82	3,435.20	2.4109	3.902
25	**51.21**	**5,045.10**	**21.6230**	**3.731**

Table 4.14 Forecasting results of ARIMA, GRNN, TF-ε-SVR-SA, SVRGASA, and SVRCGASA models (unit: hundred million kWh)

Time point (month)	Actual	ARIMA (1,1,1)	GRNN ($\sigma = 3.33$)	TF-ε-SVR-SA	SVRGASA	SVRCGASA
Oct. 2008	181.07	192.932	191.131	184.504	183.563	177.300
Nov. 2008	180.56	191.127	187.827	190.361	183.898	177.443
Dec. 2008	189.03	189.916	184.999	202.980	183.808	177.585
Jan. 2009	182.07	191.995	185.613	195.753	184.128	177.726
Feb. 2009	167.35	189.940	184.397	167.580	184.152	177.867
Mar. 2009	189.30	183.988	178.988	185.936	183.387	178.008
Apr. 2009	175.84	189.348	181.395	180.165	183.625	178.682
MAPE (%)		6.044	4.636	3.799	3.530	3.731

Similarly, in the training stage, the rolling-based forecasting procedure is also employed to obtain the forecasting load and receive training error in the training stage. Then, if training error improvement occurs, the three kernel parameters—σ, C, and ε—of the SVRCGASA model adjusted by CGASA algorithm are employed to calculate the validation error. The adjusted parameters with minimum validation error are also selected as the most appropriate parameters. The forecasting results and the suitable parameters for the SVRCGASA model are illustrated in Table 4.13, in which it is also indicated that these two models all perform the best when 25 fed-in data are used.

Table 4.14 shows the actual values and the forecast values obtained using various forecasting models: ARIMA(1,1,1), GRNN($\sigma = 3.33$), TF-ε-SVR-SA, SVRGASA, and SVRCGASA. The MAPE values are calculated to compare fairly the proposed models with other alternative models. The proposed SVRCGASA model only has smaller MAPE values than ARIMA, GRNN, TF-ε-SVR-SA models, but SVRGASA model. Furthermore, to verify the significance of accuracy improvement of SVRCGASA model comparing with ARIMA(1,1,1), GRNN ($\sigma = 3.33$), TF-ε-SVR-SA, and SVRGASA models, the Wilcoxon signed-rank test and asymptotic test, as mentioned, are also conducted. The test results are shown in Tables 4.15 and 4.16, respectively. Clearly, the SVRCGASA model is significantly superior to ARIMA(1,1,1), GRNN($\sigma = 3.33$), and TF-ε-SVR-SA models, but SVRGASA model. Meanwhile, based on Tables 4.14, 4.15, 4.16, SVRGASA model has smaller MAPE value but not complete significantly

Table 4.15 Wilcoxon signed-rank test

	Wilcoxon signed-rank test	
	$\alpha = 0.025$	$\alpha = 0.05$
Compared models	$W = 2$	$W = 3$
SVRCGASA vs. ARIMA(1,1,1)	0[a]	0[a]
SVRCGASA vs. GRNN($\sigma = 3.33$)	0[a]	0[a]
SVRCGASA vs. TF-ε-SVR-SA	1[a]	1[a]
SVRCGASA vs. SVRGASA	0[b]	0[b]

[a]Denotes that SVRCGASA model significantly outperforms other alternative models
[b]Denotes that SVRCGASA model is significantly outperformed by other alternative models

Table 4.16 Asymptotic test

	Asymptotic (S_1) test	
Compared models	$\alpha = 0.05$	$\alpha = 0.10$
SVRCGASA vs. ARIMA(1,1,1)	H_0: $e_1 = e_2$ $S_1 = -13.437$; $p = 0.000$ (reject H_0)	H_0: $e_1 = e_2$ $S_1 = -13.437$; $p = 0.000$ (reject H_0)
SVRCGASA vs. GRNN($\sigma = 3.33$)	H_0: $e_1 = e_2$ $S_1 = -2.179$; $p = 0.0146$ (reject H_0)	H_0: $e_1 = e_2$ $S_1 = -2.179$; $p = 0.0146$ (reject H_0)
SVRCGASA vs. TF-ε-SVR-SA	H_0: $e_1 = e_2$ $S_1 = -1.829$; $p = 0.0337$ (reject H_0)	H_0: $e_1 = e_2$ $S_1 = -1.829$; $p = 0.0337$ (reject H_0)
SVRCGASA vs. SVRGASA	H_0: $e_1 = e_2$ $S_1 = -0.6722$; $p = 0.2507$ (not reject H_0)	H_0: $e_1 = e_2$ $S_1 = -0.6722$; $p = 0.2507$ (not reject H_0)

(only receives significance with both levels in Wilcoxon test, but all fails with both levels in asymptotic test) outperforms SVRCGASA model. Figure 4.11 is provided to illustrate the forecasting accuracy among different models.

The reasons that SVRCGASA model cannot outperform SVRGASA model could be caused from the long hybridization process, that is, chaotic sequence to be hybridized into GA and then to be combined with SA algorithm, thus, too many transformation times to lose the searching function of evolutionary algorithm itself. Even if single chaotic sequence, GA, or SA algorithm has its superiority, after hybridization processes, however, the robust function from these technologies will be lost. For example, based on Tables 4.13 and 3.12, the CGASA algorithm is worst to shift the local solution of SVRGASA model, $(\sigma, C, \varepsilon) = (92.09, 2449.50, 13.6390)$ with local optimal forecasting errors, in terms of MAPE (3.530), to another worse solution, $(\sigma, C, \varepsilon) = (51.21, 5045.10, 21.6230)$ of SVRCGASA model to be the appropriate optimal forecasting error in terms of MAPE (3.731). Thus, it reveals that chaotic sequence is employed to avoid trapping in local minimum but wrongly to achieve worse solution of forecasting accuracy. Therefore, it should be noticed that how to avoid over-hybridizing evolutionary

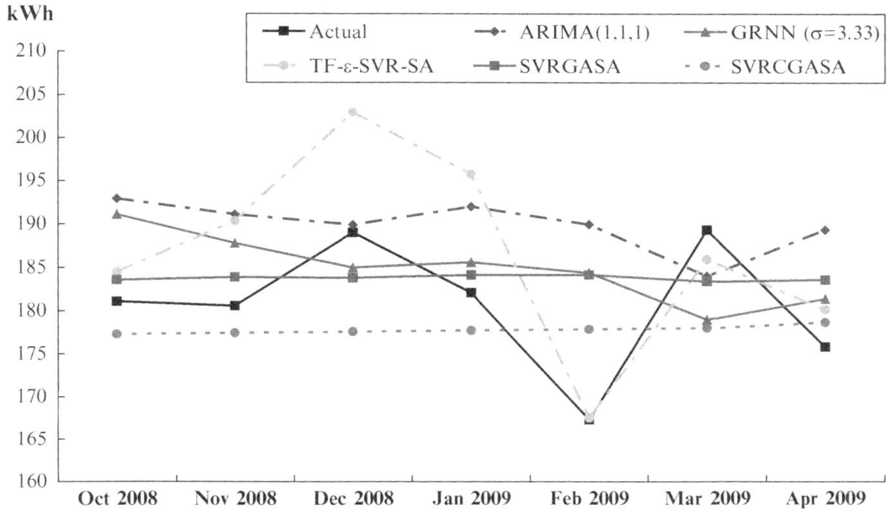

Fig. 4.11 Forecasting results of ARIMA, GRNN, TF-ε-SVR-SA, SVRGASA, and SVRCGASA models

algorithms is also an important issue while looking for more hybrid approaches. In the meanwhile, it is also clear that SVRCGASA and SVRGASA models are also not fitting the actual electric loads very well but better than other alternative models. Therefore, it also still requires hybridizing other novel mechanism to improve this shortcoming.

4.6 Chaotic PSO (CPSO) Algorithm in SVR's Parameters Determination

4.6.1 Shortcomings of PSO Algorithm and Improved by Adaptive Inertia Weight Factor and Chaotic Local Search

As mentioned in Chap. 3, Sect. 3.6, PSO algorithm has memory to store the knowledge of good solutions by all particles; in addition, particles in the swarm share information with each other. Therefore, PSO has gained much attention and wide applications in solving continuous nonlinear optimization problems. However, the performance of PSO greatly depends on its parameters, and similar to GA and SA algorithm, it often suffers from being trapped in local optimum [7, 41].

With the easy implementation and special ability to avoid being trapped in local optimum [1], chaos and chaos-based searching algorithms have aroused intense interests [7, 8]. This investigation presented in this book is motivated by a desire to improve the inefficient disadvantages of searching algorithms mentioned above in determining the three free parameters in the SVR model. Therefore, the chaotic particle swarm optimization (CPSO) algorithm proposed by Liu et al. [7] is employed in an SVR model, namely, SVRCPSO, to provide good forecasting performance in capturing nonlinear electric load changes tendency.

Based on the PSO procedure, the performance mainly depends on its parameters, and it often leads to be trapped in local optimum [41]. For example, the inertia weight l is affected by the previous velocity on the current one, a larger inertia weight pressures toward global exploration, while a smaller inertia weight pressures toward fine-tuning of the current search area. Thus, proper control of l is very important to find the optimum solution accurately. To deal with this shortcoming, Liu et al. [7] employed a chaotic PSO (CPSO) algorithm that combined PSO with adaptive inertia weight factor (AIWF) and chaotic local search (CLS). In addition, Cai et al. [8] also introduce another CPSO model, by applying other famous chaotic system, tent equation, to combine PSO with AIWF. Based on their results, there is no significant difference among these two CPSO algorithms. Due to overcoming trapped into local minimum of PSO, this investigation only applied Liu et al.'s [7] CPSO algorithm to test the potentiality in searching three parameters in an SVR model.

4.6.2 Operation Procedure of CPSO Algorithm

The operation of CPSO algorithm is based on the main procedure of PSO algorithm, that is, each particle remembers the best position it reached during the past; the local search method (via self experience) with global search methods (via neighboring experience) is employed to look for the solution. Then, adaptive inertia weight factor (AIWF) and chaotic local search (CLS) are used to play well local search.

The position, the velocity, and own best position of the ith particle pair, due to the three parameters in an SVR model, in the n-dimensional space can be represented as Eqs. (4.24)–(4.26), respectively,

$$X_{(k)i} = \left[x_{(k)i,1}, x_{(k)i,2}, \ldots, x_{(k)i,n}\right], \tag{4.24}$$

$$V_{(k)i} = \left[v_{(k)i,1}, v_{(k)i,2}, \ldots, v_{(k)i,n}\right], \tag{4.25}$$

$$P_{(k)i} = \left[p_{(k)i,1}, p_{(k)i,2}, \ldots, p_{(k)i,n}\right], \tag{4.26}$$

where $k = \sigma, C, \varepsilon$, and $i = 1, 2, \ldots, N$.

The global best position among all particles in the swarm $\mathbf{X}_{(k)i} = \left[X_{(k)1}, X_{(k)2}, \ldots, X_{(k)N}\right]$ is shown as Eq. (4.27),

$$P_{(k)g} = \left[p_{(k)g,1}, p_{(k)g,2}, \ldots, p_{(k)g,d}\right], \tag{4.27}$$

where $k = \sigma, C, \varepsilon$, and $g = 1, 2, \ldots, N$.

Then, the new velocity of each particle is computed by Eq. (4.28),

$$V_{(k)i}(t+1) = lV_{(k)i}(t) + q_1\text{rand}(\cdot)(P_{(k)i} - X_{(k)i}(t)) + q_2\text{Rand}(\cdot)(P_{(k)g} - X_{(k)i}(t)), \tag{4.28}$$

where $k = \sigma, C, \varepsilon$, and $i = 1, 2, \ldots, N$, l is called the inertia weight that controls the impact of the previous velocity of the particle on its current one; q_1 and q_2 are two positive constants called acceleration coefficients; rand(·) and Rand(·) are two independently uniformly distributed random variables with range [0,1]. AIWF is used to encourage good particles (pairs) to revise their exploration to refine results by local search and bad ones to modify searching space with large step. AIWF is determined as Eq. (4.29).

$$l = \begin{cases} l_{\min} + \frac{(l_{\max} - l_{\min})(f_i - f_{\min})}{f_{\text{avg}} - f_{\min}}, & f_i \leq f_{\text{avg}} \\ l_{\max}, & f_i \geq f_{\text{avg}} \end{cases}, \tag{4.29}$$

where $l_{\max}$ and $l_{\min}$ are the maximum and minimum of l, respectively. f_i is the current objective value of ith particle pair, f_{avg} and $f_{\min}$ are the average and minimum objective values of all particle pairs, respectively.

After the velocity has been updated, the new position of the particle for each parameter in the next generation is determined as Eq. (4.30),

$$X_{(k)i}(t+1) = X_{(k)i}(t) + V_{(k)i}(t+1), \tag{4.30}$$

where $k = \sigma, C, \varepsilon$, and $i = 1, 2, \ldots, N$.

Notice that the value of each component in $V_{(k)i}$ can be limited to the range $[-v_{\max}, v_{\max}]$ to control excessive roaming of particles outside the search space. This process is repeated until the defined stopping threshold is reached. The procedure of CPSO algorithm is illustrated as follows and the flowchart is shown as Fig. 4.12. The CPSO algorithm is used to seek a better combination of the three parameters in the SVR so that a smaller MAPE is obtained during forecasting iteration.

Step 1: Initialization. Initialize a defined population of particle pairs (σ_i, C_i, ε_i) with random positions ($X_{\sigma i}$, X_{Ci}, $X_{\varepsilon i}$) and velocities ($V_{\sigma i}$, V_{Ci}, $V_{\varepsilon i}$), where each particle contains n variables.

Step 2: Objective value computing. Compute the objective values (forecasting errors) of all particle pairs. Let own best position ($P_{\sigma i}$, P_{Ci}, $P_{\varepsilon i}$) of each particle pair and its objective value $f_{\text{best}i}$ equal to its initial position and objective value. Let global best position ($P_{\sigma g}$, P_{Cg}, $P_{\varepsilon g}$) and its objective value $f_{\text{globalbest}i}$ equal to the best initial particle pair's position and its objective value.

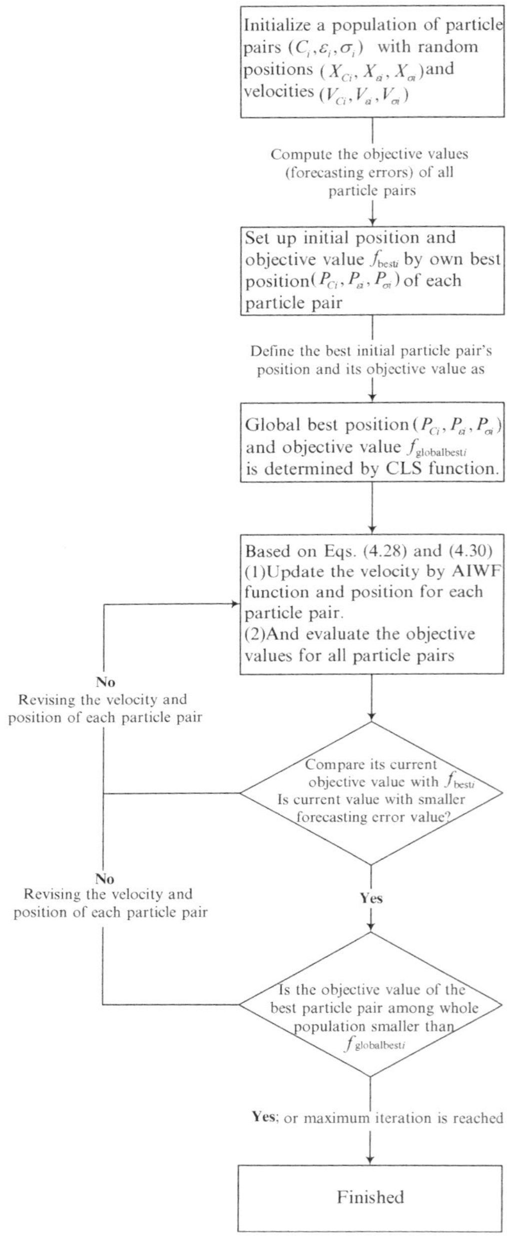

Fig. 4.12 The architecture of chaotic particle swarm optimization (CPSO) algorithm

Step 3: Evaluation of the objective values. According to Eqs. (4.28) and (4.29), update the velocity and position for each particle pair and evaluate the objective values for all particle pairs.

Step 4: Comparison and update. For each particle pair, compare its current objective value with $f_{\text{best}i}$. If the current objective value is better (i.e., with smaller forecasting accuracy index value), then update best position ($P_{\sigma i}$, P_{Ci}, $P_{\varepsilon i}$) and its objective value by the current position and objective value.

Step 5: The best particle pair determination. Determine the best particle pair of whole population based on the best objective value. CLS is used to perform locally oriented search (exploitation) for the solution $f_{\text{globalbest}i}$; the detail procedure is illustrated as below paragraph. If the objective value is smaller than $f_{\text{globalbest}i}$, then update ($P_{\sigma g}$, P_{Cg}, $P_{\varepsilon g}$) and its objective value with the current best particle pair's position and objective.

Step 6: Stop criterion. If a stopping threshold (forecasting accuracy) is reached, then ($P_{\sigma g}$, P_{Cg}, $P_{\varepsilon g}$) and its $f_{\text{globalbest}i}$ would be determined; otherwise, go back to step 3.

CLS is used to perform locally oriented search (exploitation) for the solution $f_{\text{globalbest}i}$, which resulted from PSO algorithm. CLS is based on the logistic equation, [see Eq. (4.5)] with sensitive dependence on initial conditions. The procedure of CLS is illustrated as follows:

Step 1. Setting $\lambda = 0$ and employing Eq. (4.31) to map the three parameters, $X_{(k)i}^{(\lambda)}$, $k = \sigma, C, \varepsilon$, $i = 1, 2, \ldots, N$, among the intervals ($x_{\min(k)i}$, $x_{\max(k)i}$) into chaotic variable $cx_{(k)i}^{(\lambda)}$ located in the interval (0,1).

$$cx_{(k)i}^{(\lambda)} = \frac{X_{(k)i}^{(\lambda)} - x_{\min(k)i}}{x_{\max\,(k)i} - x_{\min(k)i}} \tag{4.31}$$

Step 2. Use Eq. (4.5) to compute the next iteration chaotic variable $cx_{(k)i}^{(\lambda+1)}$.

Step 3. Transform $cx_{(k)i}^{(\lambda+1)}$ to obtain three parameters for the next iteration, $X_{(k)i}^{(\lambda+1)}$, by the following Eq. (4.32):

$$X_{(k)i}^{(\lambda+1)} = x_{\min\,(k)i} + cx_{(k)i}^{(\lambda+1)} \left(x_{\max\,(k)i} - x_{\min(k)i}\right). \tag{4.32}$$

Step 4. Compute the new objective value with $X_{(k)i}^{(\lambda+1)}$.

Step 5. If the new objective value with smaller forecasting accuracy index value or maximum iteration is reached, then the new chaotic variable $X_{(k)i}^{(\lambda+1)}$ and its corresponding objective value is the final solution; otherwise, let $\lambda = \lambda + 1$ and go back to step 2.

In the investigation, the mean absolute percentage error (MAPE), shown as Eq. (3.13), serves as the forecasting accuracy index for identifying suitable parameters, determined in step 4 of CPSO algorithm and in step 5 of CLS.

Table 4.17 Parameters setting of PSO algorithm

Population size	Maximal iteration	Velocity limit						Inertia weight l		Acceleration coefficients					
		$-v_{max}$			v_{max}			l_{min}	l_{max}	q_1			q_2		
		σ	C	ε	σ	C	ε			σ	C	ε	σ	C	ε
20	10,000	−0.5	−2,500	−15	0.5	2,500	15	0.2	1.2	0.05	100	0.5	0.05	100	0.5

Table 4.18 Parameters determination of SVRCPSO model

Nos. of fed-in data	Parameters			MAPE of testing (%)
	σ	C	ε	
5	88.02	445.99	2.1751	4.502
10	11.74	180.91	0.6728	4.079
15	4.60	3,886.10	10.1500	3.716
20	356.64	4,433.30	13.6800	3.502
25	**433.23**	**9,855.20**	**2.1174**	**3.231**

4.6.3 CPSO Algorithm for Three-Parameter Determination and Forecasting Results

This subsection will demonstrate the three-parameter determination of the proposed hybrid model (SVR with CPSO), namely, SVRCPSO model.

The parameters of the CPSO algorithm in the proposed model are experimentally set as shown in Table 4.17. The population size is set as 20; the total number of function evaluation is fixed as 10,000; q_1 and q_2 for each particle pair (σ, C, ε) are set to 0.05, 100, 0.5, respectively. v_{max} for σ particle are both clamped to be 10 % of its search space (where $\sigma \in [0,500]$). v_{max} for C particle is clamped to be 12.5 % of its search space ($C \in [0,20000]$). v_{max} for ε particle are both clamped to be 15 % of its search space ($\varepsilon \in [0,100]$). The standard PSO algorithm [42] uses a linearly varying inertia weight over the generations, varying from 1.2 at the beginning of the search to 0.2 at the end. The CPSO algorithm uses the AIWF defined in Eq. (4.29) with $l_{max} = 1.2$ and $l_{min} = 0.2$.

Similarly, in the training stage, the rolling-based forecasting procedure is also employed to obtain the forecasting load and receive training error in the training stage. Then, if training errors improvement occurs, the three kernel parameters—σ, C, and ε—of the SVRCPSO model adjusted by CPSO algorithm are employed to calculate the validation error. The adjusted parameters with minimum validation error are also selected as the most appropriate parameters. The forecasting results and the suitable parameters for the SVRCPSO model are illustrated in Table 4.18, in which it is also indicated that these two models all perform the best when 25 fed-in data are used.

Table 4.19 shows the actual values and the forecast values obtained using various forecasting models: ARIMA(1,1,1), GRNN($\sigma = 3.33$), TF-ε-SVR-SA, SVRPSO, and SVRCPSO. The MAPE values are calculated to compare fairly the

Table 4.19 Forecasting results of ARIMA, GRNN, TF-ε-SVR-SA, SVRPSO, and SVRCPSO models (unit: hundred million kWh)

Time point (month)	Actual	ARIMA(1,1,1)	GRNN(σ = 3.33)	TF-ε-SVR-SA	SVRPSO	SVRCPSO
Oct. 2008	181.07	192.932	191.131	184.504	184.042	181.938
Nov. 2008	180.56	191.127	187.827	190.361	183.577	182.186
Dec. 2008	189.03	189.916	184.999	202.980	183.471	182.677
Jan. 2009	182.07	191.995	185.613	195.753	184.210	182.794
Feb. 2009	167.35	189.940	184.397	167.580	184.338	182.826
Mar. 2009	189.30	183.988	178.988	185.936	183.725	182.746
Apr. 2009	175.84	189.348	181.395	180.165	184.529	184.222
MAPE (%)		6.044	4.636	3.799	3.638	3.231

Table 4.20 Wilcoxon signed-rank test

	Wilcoxon signed-rank test	
	$\alpha = 0.025$	$\alpha = 0.05$
Compared models	$W = 2$	$W = 3$
SVRCPSO vs. ARIMA(1,1,1)	0[a]	0[a]
SVRCPSO vs. GRNN(σ = 3.33)	2[a]	2[a]
SVRCPSO vs. TF-ε-SVR-SA	2[a]	2[a]
SVRCPSO vs. SVRPSO	0[a]	0[a]

[a]Denote that SVRCPSO model significantly outperforms other alternative models

proposed models with other alternative models. The proposed SVRCPSO model has smaller MAPE values than ARIMA, GRNN, TF-ε-SVR-SA, and SVRPSO model. Furthermore, to verify the significance of accuracy improvement of SVRCPSO model comparing with ARIMA(1,1,1), GRNN(σ = 3.33), TF-ε-SVR-SA, and SVRPSO models, the Wilcoxon signed-rank test and asymptotic test, as mentioned, are also conducted. The test results are shown in Tables 4.20 and 4.21, respectively. Clearly, the SVRCPSO model is significantly superior to ARIMA (1,1,1), GRNN(σ = 3.33), TF-ε-SVR-SA, and SVRPSO models. Figure 4.13 is provided to illustrate the forecasting accuracy among different models.

The superior performance of the SVRCPSO model is not only caused from several similar reasons of SVRPSO model, such as SVR-based model with nonlinear mapping capabilities, minimizing the structural risks rather than the training errors and the quadratic programming technique which is based on the assumptions of convex set and existence of global optimum solution, but also is caused from the chaotic local search (CLS) and adaptive inertia weight factor (AIWF) of CPSO algorithm. Therefore, it will be theoretically approximated to the global optimum solution if the superior searching algorithm is used. For example, based on Tables 4.18 and 3.17, the CPSO algorithm is excellently to shift the local solution of SVRPSO model, $(\sigma, C, \varepsilon) = (158.44, 7014.50, 2.2836)$ with local optimal forecasting errors, in terms of MAPE (3.638), to another better solution, $(\sigma, C, \varepsilon) = (433.23, 9855.20, 2.1174)$ of SVRCPSO model to be the appropriate optimal forecasting error in terms

Table 4.21 Asymptotic test

Compared models	Asymptotic (S_1) test $\alpha = 0.05$	$\alpha = 0.10$
SVRCPSO vs. ARIMA(1,1,1)	H_0: $e_1 = e_2$ $S_1 = -10.476$; $p = 0.000$ (reject H_0)	H_0: $e_1 = e_2$ $S_1 = -10.476$; $p = 0.000$ (reject H_0)
SVRCPSO vs. GRNN($\sigma = 3.33$)	H_0: $e_1 = e_2$ $S_1 = -2.066$; $p = 0.0194$ (reject H_0)	H_0: $e_1 = e_2$ $S_1 = -2.066$; $p = 0.0194$ (reject H_0)
SVRCPSO vs. TF-ε-SVR-SA	H_0: $e_1 = e_2$ $S_1 = -3.377$; $p = 0.00036$ (reject H_0)	H_0: $e_1 = e_2$ $S_1 = -3.377$; $p = 0.00036$ (reject H_0)
SVRCPSO vs. SVRPSO	H_0: $e_1 = e_2$ $S_1 = -49.021$; $p = 0.000$ (reject H_0)	H_0: $e_1 = e_2$ $S_1 = -49.021$; $p = 0.000$ (reject H_0)

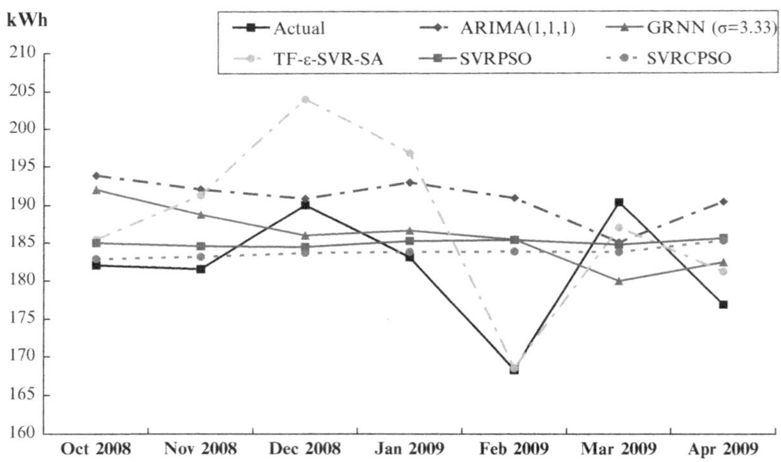

Fig. 4.13 Forecasting results of ARIMA, GRNN, TF-ε-SVR-SA, SVRPSO, and SVRCPSO models

of MAPE (3.231). Thus, it reveals that CLS and AIWF are capable to avoid trapping in local minimum to achieve forecasting accuracy improvement by hybridized into the SVR model. In the future research, it not only should be focused on the feasibility of hybrid evolutionary algorithms to adjust the parameters of SVR-based forecasting models but also should be focused on how to well determine the interior parameters of evolutionary algorithms. For example, in this section, CLS is hybridized with PSO algorithm to adjust the interior results from PSO to achieve outstanding forecasting performance; moreover, the parameters of PSO (such as the velocity $v_{\max}$, two positive acceleration coefficients, q_1 and q_2) and of AIWF (the maximum $l_{\max}$ and minimum $l_{\min}$ of inertia weight l) should be further determined by other evolutionary algorithms to obtain more appropriate parameter combination and to achieve more improvable, satisfactory, and accurate electric load forecasting if it exists.

Finally, however, it is also clear that SVRCPSO model is also not fitting the actual electric loads very well but better than other alternative models, even it has significant smaller MAPE values than other alternatives. Therefore, it also still requires hybridizing other novel mechanism to improve this shortcoming.

4.7 Chaotic Ant Swarm Optimization Algorithm in SVR's Parameters Determination

4.7.1 Shortcomings of CACO Algorithm and Improved by Chaotic Sequence

In 1992, Marco Dorigo first proposed ant colony optimization (ACO) inspired by the behavior of ant colonies [43, 44]. In ACO, every artificial ant colony cooperates to forage for food, wherein each ant lays down the pheromone on the path to the food sources or back to the nest. The pheromone trail is the most important process for individual ant to determine its route. It could be ensured global optimum or near-optimum of a function in a search space by implementing heuristic adjustments for the probability of an ant moving paths combination. The probability of an ant for a path to be followed is based on its pheromone amount (pheromone trail is roaming randomly), that is, random meta-heuristic of nondeterministic probability theory [45]. However, ACO is originally proposed for discrete optimization and their application to continuous optimization problems require some specified transformation techniques. With the easy implementation and special ability to avoid being trapped in local optimum [1], chaos and chaos-based searching algorithms have aroused intense interests [7, 8]. Cole [46] indicates that "variation in the temporal component of individual behavior may not be due simply to chance variations in the stochastic world, but to deterministic processes that depend on initial conditions," and then, he concludes that ant colony exhibits a periodic behavior while single ant shows low-dimensional deterministic chaotic activity patterns. Thus, combining the chaotic behaviors of individual ant with the intelligent organization foraging actions of ant colony is proposed, namely, chaotic ant swarm optimization (CAS) algorithm [47]. This section is motivated by a desire to improve the inefficient disadvantages of searching algorithms mentioned above in determining the three free parameters in the SVR model. The CAS algorithm proposed by Li et al. [47] is employed, namely, SVRCAS, to provide good forecasting performance in capturing nonlinear electric load changes tendency.

4.7.2 Operation Procedure of CAS Algorithm

In the CAS, the organization variable is employed to perform self-organization foraging process of ant colony. During the best food source searching process, the organization variable will increase its influences on the individual ant and associate ant colony. First, the affect of the organization variable on individual ant is too weak to let the search behavior of an individual ant be chaotic, that is, this phase reveals non-coordinated among ants. Second, evolving in time and space, the organization will continually increase its influences on the behavior of individual ant, that is, the chaotic behavior of the individual decreases gradually and coordination among ants (ant colony) phase starts. Via the communication of previously best positions with neighbors, the single ant adjusts his position and moves to the best one, which has ever been found in the search space.

To obtain the chaotic search initially, chaotic system, represented by the famous logistic function, as Eq. (4.5), is employed. The adjustment of the chaotic behavior of individual ant is achieved by the organization variable y_i, and y_i will lead the individual ant to move to the new site acquired with the best fitness value eventually as following Eq. (4.33) [45]:

$$\begin{cases} y_i^{(n)} = \left(y_i^{(n-1)}\right)^{1+r_i} \\ cx_{(k)id}^{(n)} = \left(cx_{(k)id}^{(n-1)} + v_i\frac{7.5}{\psi_d}\right)\exp\left(\left(1-\exp(-ay_i^{(n)})\right)\left(3-\psi_d cx_{(k)id}^{(n-1)} + v_i\frac{7.5}{\psi_d}\right)\right) \\ \qquad -v_i\frac{7.5}{\psi_d} + (f\text{best}_{(k)id}^{(n-1)} - cx_{(k)id}^{(n-1)})\exp(-2ay_i^{(n)} + b) \end{cases} \tag{4.33}$$

where $cx_{(k)id}^{(n)}$ is the current state of the dth dimension of the individual ant i, where $d = 1, 2, \ldots, l$, l is the dimensional numbers of search space; $cx_{(k)id}^{(1)} = \frac{7.5}{\psi_d} \times (1 - v_i) \times$ rand(1), where rand(1) is a uniformly distributed random number in (0,1); $f\text{best}_{(k)id}^{(n)}$ is the best position found by the ith ant and its neighbors within n iterations; v_i determines the search region of ant i, to be simplified, set $0 < v_i < 1$; a is a sufficiently large positive constant and can be selected as $a = 200$; b is a constant, $0 \leq b \leq 2/3$; r_i is the organization factor of ant i, which affects the convergence speed of the CAS algorithm directly, if r_i is very large, it implies the iteration step of "chaotic" search is small; thus, the system converges quickly and the desired optima or near-optima cannot be achieved, vice versa, the value of r_i is chosen typically as $0 \leq r_i \leq 0.5$, each ant could have different r_i, such as $r_i = 0.3 + 0.02 \times \text{rand}(1)$. ψ_d affects the search ranges of the CAS algorithm, if the interval of the search is ω_d, then, $\psi_d \approx 7.5/\omega_d$.

In the CAS algorithm, the neighbors of the ant will be defined finite ants based on their distance in space. Generally, the neighbor selection can be defined as the following two ways. The first is the nearest fixed number of neighbors, the nearest m ants are defined as the neighbors of single ith ant. However, this way could not reveal the affects of self-organization behaviors of ants. The second way is, due to the influence of self-organization behaviors of ants, to consider the situation in which the number of neighbors increasing with iterative steps. As mentioned above, the impact of organization will become stronger and the neighbors of the ant will increase to coordinate with each other and to adjust positions to the best one, thus, the number of nearest neighbors will be dynamically changed as time evolves or iterative steps increase. On the other hand, when every individual trajectory is adjusted toward the successes of neighbors, the swarm will converge in optimal regions of the search space. The search of some ants will fail if the individual cannot obtain information about the best food source from their neighbors. To simulate the way of dynamical neighbors, the number q of single ant is defined to increase for every T iterations. In addition, to determine the candidate neighbors, the Euclidian distance is employed to calculate the distance between the two ants as Eq. (4.34),

$$\sqrt{\left(cx_{(k)i1}^{(n)} - cx_{(k)j1}^{(n)}\right)^2 + \left(cx_{(k)i2}^{(n)} - cx_{(k)j2}^{(n)}\right)^2 + \ldots + \left(cx_{(k)il}^{(n)} - cx_{(k)jl}^{(n)}\right)^2}, \tag{4.34}$$

where $cx_{(k)id}^{(n)} = (cx_{(k)i1}^{(n)}, cx_{(k)i2}^{(n)}, \ldots, cx_{(k)il}^{(n)})$ and $cx_{(k)jd}^{(n)} = (cx_{(k)j1}^{(n)}, cx_{(k)j2}^{(n)}, \ldots, cx_{(k)jl}^{(n)})$ are defined as the positions of two ants (the ith and jth ant), respectively, $i, j = 1, 2, \ldots, N_k$, $i \neq j$.

The proposed procedure of CAS algorithm is illustrated as follows and the flowchart is shown as Fig. 4.14. The relevant parameters settings of CAS algorithm are listed in Table 4.22.

Step 1: Initialization. Define three ant colonies, namely, σ-ant colony, C-ant colony, and ε-ant colony, to represent the three SVR positive parameters, σ, C, and ε, respectively. The numbers of ants are set to be 10 in each ant colony searching, that is, totally 30 ants for each iteration searching. The maximum number of iterations is set to 20,000 to avoid infinite iterations. Setting $n = 0$ and employing Eq. (4.35) to map the three parameters $X_{(k)id}^{(n)}$, $k = \sigma, C, \varepsilon$, $i = 1, 2, \ldots, N_k$ among the intervals $(x_{\min(k)i}, x_{\max(k)i})$ into chaotic variable $cx_{(k)id}^{(n)}$ located in the interval (0,1).

$$cx_{(k)id}^{(n)} = \frac{X_{(k)id}^{(n)} - x_{\min(k)i}}{x_{\max(k)i} - x_{\min(k)i}}, \quad i = 1, 2, \ldots, N_k \tag{4.35}$$

In this section, the search intervals for each ant colony are $\sigma \in (0{,}500)$, $C \in (0{,}20000)$, and $\varepsilon \in (0{,}100)$, respectively. Then, use Eq. (4.5) to compute the current chaotic variable $cx_{(k)id}^{(n)}$. In the meanwhile, also set the initial organization variable $y_i^{(1)}$ as 0.999 [47].

Step 2: Determine the candidate neighbors. By using Eq. (4.34) to calculate the Euclidian distance between each ant and its neighbors to obtain the nearest candidate neighbor, $f\text{best}_{(k)id}^{(n)}$. In this chapter, it is defined that the number of ant neighbors increases by 1 every 5 iterations with a given initial number 2 of neighbors.

Fig. 4.14 The architecture of chaotic ant swarm optimization (CAS) algorithm

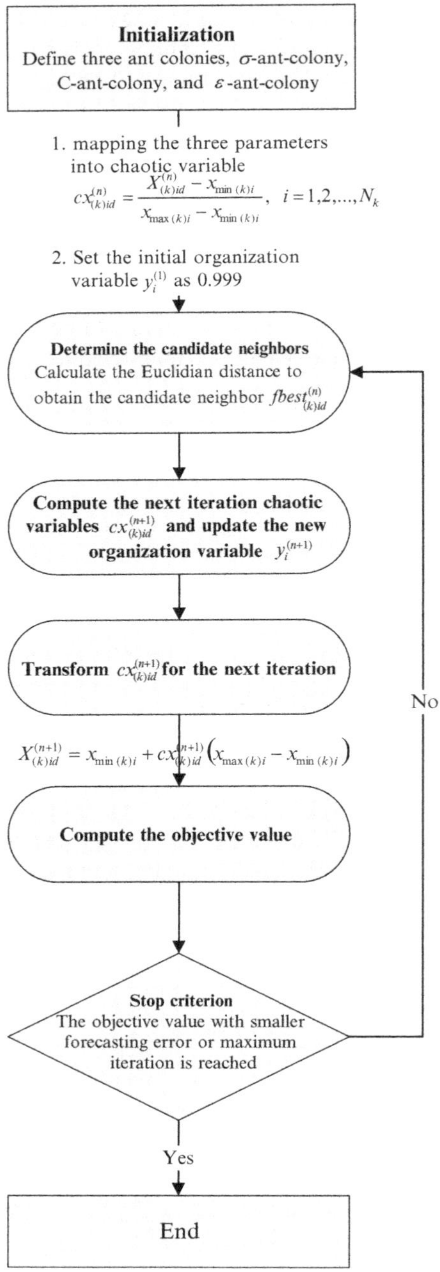

Table 4.22 CAS's parameters setting

Number of ants for each colony (N_k)	Maximal iterations (n)	Bifurcation parameter (μ)	The search region of ant i (v_i)
10	500	4	0.5
Two positive constant (a, b)	Initial organization variable ($y_i^{(1)}$)	The organization factor of ant i (r_i)	
$a = 200$	0.999	$0 \leq r_i \leq 0.5$	
$b = 0.4$		$r_i = 0.3 + 0.02 \times \text{rand}(1)$	
The search interval for σ-ant colony ($x_{\min(\sigma)i}$, $x_{\max(\sigma)i}$)	The search interval for C-ant colony ($x_{\min(C)i}$, $x_{\max(C)i}$)	The search interval for ε-ant colony ($x_{\min(\varepsilon)i}$, $x_{\max(\varepsilon)i}$)	
(0,500)	(0,20000)	(0,100)	

Step 3: Compute the next iteration chaotic variables and update the new organization variable. Use Eq. (4.33) to compute the next iteration chaotic variable, $cx_{(k)id}^{(n+1)}$, and the organization variable $y_i^{(n+1)}$.

Step 4: Transform $cx_{(k)id}^{(n+1)}$ for the next iteration. Transform $cx_{(k)id}^{(n+1)}$ to obtain three parameters for the next iteration $X_{(k)id}^{(n+1)}$, by the following Eq. (4.36):

$$X_{(k)id}^{(n+1)} = x_{\min(k)i} + cx_{(k)id}^{(n+1)} \left(x_{\max(k)i} - x_{\min(k)i}\right) \tag{4.36}$$

Step 5: Compute the objective value. Put the three parameter values ($X_{(k)id}^{(n+1)}$, $k = \sigma, C, \varepsilon, i = 1, 2, \ldots, N_k$) into the SVR model and calculate the forecasting error (i.e., objective function). In the investigation, the mean absolute percentage error (MAPE), shown as Eq. (3.13), serves as the forecasting accuracy index for identifying suitable parameters in the SVR model. The CAS algorithm is used to seek a better combination of the three parameters in the SVR so that a smaller MAPE is obtained during forecasting iteration.

Step 6: Stop criterion. If the new objective value with smaller forecasting accuracy index value or maximum iteration is reached, then the new chaotic variable $cx_{(k)id}^{(n+1)}$, the new three parameters $X_{(k)id}^{(n+1)}$, and its corresponding objective values are the final solutions; otherwise, go to the next iteration and go back to step 2.

4.7.3 CAS Algorithm for Three Parameters Determination and Forecasting Results

This subsection will demonstrate the three parameters determination of the proposed hybrid model (SVR with CAS), namely, SVRCAS model.

Similarly, in the training stage, the rolling-based forecasting procedure is also employed to obtain the forecasting load and receive training error in the training

Table 4.23 Parameters determination of SVRCAS model

Nos. of fed-in data	Parameters			MAPE of testing (%)
	σ	C	ε	
5	12.52	460.64	2.2106	4.578
10	117.24	40.795	0.1653	3.732
15	800.53	3,130.60	14.5430	3.460
20	209.30	3,780.80	16.6730	3.234
25	**50.72**	**9,589.70**	**19.5440**	**2.881**

Table 4.24 Forecasting results of ARIMA, GRNN, TF-ε-SVR-SA, SVRCACO, and SVRCAS models (unit: hundred million kWh)

Time point (month)	Actual	ARIMA (1,1,1)	GRNN ($\sigma = 3.33$)	TF-ε-SVR-SA	SVRCACO	SVRCAS
Oct. 2008	181.07	192.932	191.131	184.504	180.876	180.619
Nov. 2008	180.56	191.127	187.827	190.361	182.122	180.899
Dec. 2008	189.03	189.916	184.999	202.980	184.610	181.178
Jan. 2009	182.07	191.995	185.613	195.753	185.233	181.457
Feb. 2009	167.35	189.940	184.397	167.580	185.274	181.735
Mar. 2009	189.30	183.988	178.988	185.936	184.247	182.013
Apr. 2009	175.84	189.348	181.395	180.165	184.930	180.758
MAPE (%)		6.044	4.636	3.799	3.371	2.881

stage. Then, if training errors improvement occurs, the three kernel parameters—σ, C, and ε—of the SVRCAS model adjusted by CAS algorithm are employed to calculate the validation error. The adjusted parameters with minimum validation error are also selected as the most appropriate parameters. The forecasting results and the suitable parameters for the SVRCAS model are illustrated in Table 4.23, in which it is also indicated that these two models all perform the best when 25 fed-in data are used.

Table 4.24 shows the actual values and the forecast values obtained using various forecasting models: ARIMA(1,1,1), GRNN($\sigma = 3.33$), TF-ε-SVR-SA, SVRCACO, and SVRCAS. The MAPE values are calculated to compare fairly the proposed models with other alternative models. The proposed SVRCAS model has smaller MAPE values than ARIMA, GRNN, TF-ε-SVR-SA, and SVRCACO models. Furthermore, to verify the significance of accuracy improvement of SVRCAS model comparing with ARIMA(1,1,1), GRNN($\sigma = 3.33$), TF-ε-SVR-SA, and SVRCACO models, the Wilcoxon signed-rank test and asymptotic test, as mentioned, are also conducted. The test results are shown in Tables 4.25 and 4.26, respectively. Clearly, the SVRCAS model is significantly superior to ARIMA (1,1,1), GRNN($\sigma = 3.33$), TF-ε-SVR-SA, and SVRCACO models. Figure 4.15 is provided to illustrate the forecasting accuracy among different models.

The superior performance of the SVRCAS model is not only caused from several similar reasons of SVRCACO model, such as SVR-based model with nonlinear mapping capabilities, minimizing the structural risks rather than the training errors

Table 4.25 Wilcoxon signed-rank test

	Wilcoxon signed-rank test	
	$\alpha = 0.025$	$\alpha = 0.05$
Compared models	$W = 2$	$W = 3$
SVRCAS vs. ARIMA(1,1,1)	0[a]	0[a]
SVRCAS vs. GRNN($\sigma = 3.33$)	1[a]	1[a]
SVRCAS vs. TF-ε-SVR-SA	2[a]	2[a]
SVRCAS vs. SVRCACO	0[a]	0[a]

[a]Denote that SVRCAS model significantly outperforms other alternative models

Table 4.26 Asymptotic test

	Asymptotic (S_1) test	
Compared models	$\alpha = 0.05$	$\alpha = 0.10$
SVRCAS vs. ARIMA(1,1,1)	H_0: $e_1 = e_2$ $S_1 = -13.977$; $p = 0.000$ (reject H_0)	H_0: $e_1 = e_2$ $S_1 = -5.780$; $p = 0.000$ (reject H_0)
SVRCAS vs. GRNN($\sigma = 3.33$)	H_0: $e_1 = e_2$ $S_1 = -2.905$; $p = 0.0018$ (reject H_0)	H_0: $e_1 = e_2$ $S_1 = -2.905$; $p = 0.0018$ (reject H_0)
SVRCAS vs. TF-ε-SVR-SA	H_0: $e_1 = e_2$ $S_1 = -3.550$; $p = 0.00019$ (reject H_0)	H_0: $e_1 = e_2$ $S_1 = -3.550$; $p = 0.00019$ (reject H_0)
SVRCAS vs. SVRCACO	H_0: $e_1 = e_2$ $S_1 = -3.565$; $p = 0.00019$ (reject H_0)	H_0: $e_1 = e_2$ $S_1 = -3.565$; $p = 0.00019$ (reject H_0)

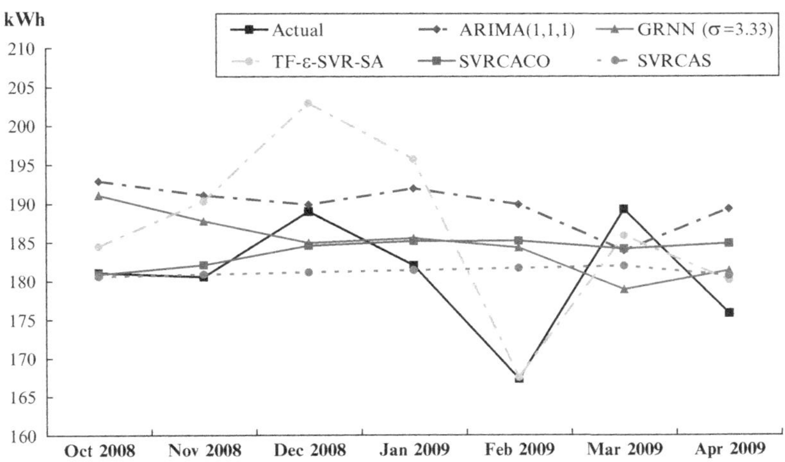

Fig. 4.15 Forecasting results of ARIMA, GRNN, TF-ε-SVR-SA, SVRCACO, and SVRCAS models

and the quadratic programming technique which is based on the assumptions of convex set and existence of global optimum solution, but also is caused from that the CAS algorithm employs the organization variable to perform self-organization foraging process of ant colony to determine proper parameters combination and applies ergodicity property of chaotic sequences to enrich the searching behavior to avoid premature convergence. Therefore, it will be theoretically approximated to the global optimum solution if the superior searching algorithm is used. For example, based on Tables 4.23 and 3.21, the CAS algorithm is excellently to shift the local solution of SVRCACO model, $(\sigma, C, \varepsilon) = (243.55, 6868.10, 11.2480)$ with local optimal forecasting errors, in terms of MAPE (3.371), to another better solution, $(\sigma, C, \varepsilon) = (50.72, 9589.70, 19.5440)$ of SVRCAS model to be the appropriate optimal forecasting error in terms of MAPE (2.881). Thus, it reveals that the self-organization foraging mechanism from the organization variable is capable to avoid trapping in local minimum to achieve forecasting accuracy improvement by hybridized into the CAS algorithm. It is interesting to address the SVRCAS model focuses on the interactions among individual ant's chaotic behavior and ant colony organization foraging activities, instead of "expert rules," to negotiate and to coordinate to look for much better solutions. Therefore, the better solution is evolved with "learning by doing" activities among ants and their colony to approximately reach the global optimum or near-optimum. It sometimes likes a "practical (numerical) rule" to guide colony organization variable enhance its affects on individual ant's chaotic behaviors. Finally, however, it is also clear that SVRCAS model is also not fitting the actual electric loads very well but better than other alternative models, even it has significant smaller MAPE values than other alternatives. Therefore, it also still requires hybridizing other novel mechanism to improve this shortcoming.

4.8 Chaotic Artificial Bee Colony Algorithm in SVR's Parameters Determination

4.8.1 Shortcomings of Chaotic Artificial Bee Colony Algorithm and Improved by Chaotic Sequence

As known that lots of optimization algorithms conduct only one search operation during one iteration time, for example, PSO carries out global search at the beginning and local search in the later stage. For ABC algorithm, it conducts both global search and local search during one iteration time to ensure that ABC algorithm is more probabilistic to receive more suitable parameters combination, and thereby, efficiently to avoid local optimum to a large extent. However, while the solution domain of the problem occurs in a region, such as a plateau with three dimensions, where all the surrounding potential solutions are closed to the same fitness, if the ABC is employed to find out the best solution, it will cost large number of iterations to reach the global optimal solution. In addition, if lots of multiple peaks are proximately closed to the same fitness and there are very deep divides among them, then, the ABC algorithm will be troubled in "hopping" from one peak to another peak [48]. Eventually, after set (or limited) iterations, it will converge prematurely. To overcome the shortcoming, it is necessary to find some effective approaches (i.e., some improved designs or procedures on ABC algorithm) to track in the solution space effectively and efficiently. One feasible approach is focused on the chaos approach, due to its easy implementation and special ability to avoid being trapped in local optimum [1]. The application of chaotic sequences can be a good alternative to diversify the initial definition domain in stochastic optimization procedures, that is, small changes in the parameter settings or the initial values in the model. After the search process of each bee, the chaotic search in the neighborhood of the current best solution is conducted to choose one better solution into the next generation. Due to the ergodicity property of chaotic sequences, it will lead to very different future solution-finding behaviors; thus, chaotic sequences can be used to enrich the searching behavior and to avoid being trapped into local optimum [49].

4.8.2 Operation Procedure of Chaotic Artificial Bee Colony Algorithm

The proposed procedure of Chaotic artificial bee colony (CABC) is illustrated as follows and the flowchart is shown as Fig. 4.16.

Step 1: Initialization. Initialize the population size N_{p}, the number of employed foragers n_{e}, and the number of unemployed foragers (onlookers) n_{O}, which satisfy the condition, $N_{\mathrm{p}} = n_{\mathrm{e}} + n_{\mathrm{O}}$. Let x_{ij} ($i = 1, 2, \ldots, N_{\mathrm{p}}$; $j = 1, 2, \ldots, D$) represent the initial solution of parameter combination in an SVR model, where D is the number of parameters. D is set as 3 in this section.

Step 2: Criteria of food source determination. Based on ABC, choosing a food source of an onlooker is depended on the probability value associated with that food source. However, for forecasting accuracy improvement in the investigation, the onlooker will choose a food source according to the mean absolute percentage error (MAPE), shown as Eq. (3.13), which is also served as the forecasting accuracy index.

Step 3: Generate neighboring food sources (potential solutions). A candidate food source (solution) v_{ij} from the old solution x_{ij} can be generated as Eq. (4.37):

$$v_{ij} = x_{ij} + \Phi_{ij}\left(x_{ij} - x_{kj}\right), \tag{4.37}$$

where $k \in \{1, 2, \ldots, N\}$ are randomly chosen indices, k must be different from i, and Φ_{ij} is a random number in the range $[-1,1]$. If the MAPE value of the candidate solution v_{ij} is equal to or smaller than x_{ij}'s, then v_{ij} will be set as the new solution; else, x_{ij} will remain as the solution.

Fig. 4.16 The architecture of chaotic artificial bee colony (CABC) algorithm

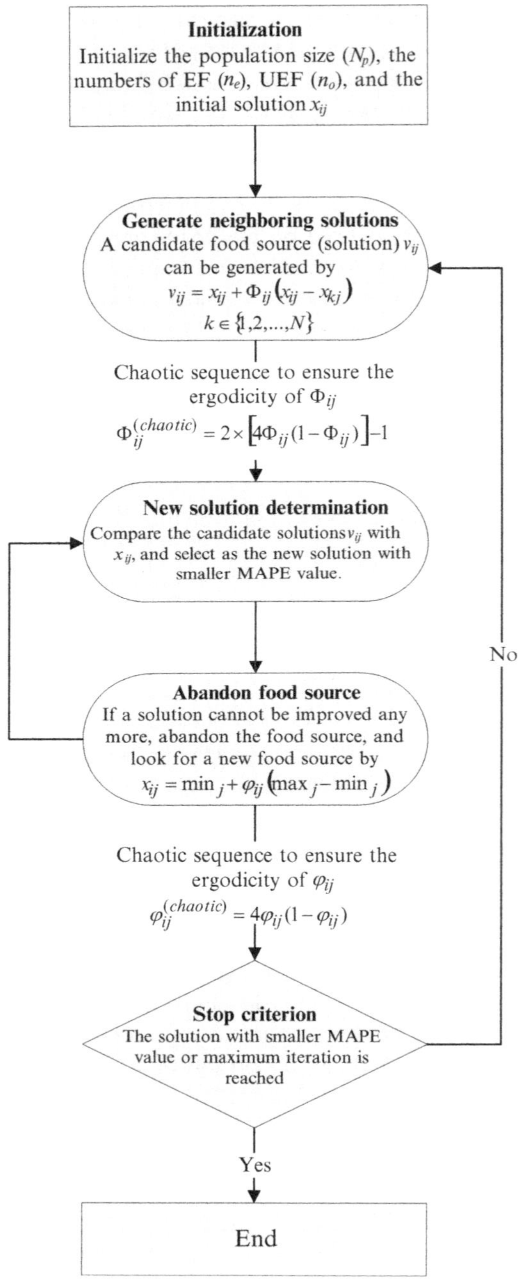

The parameter Φ_{ij} in ABC is the key factor that affects convergence [50]. However, Eq. (4.37) is pseudo-randomly generated but it cannot ensure if the random variable can travel ergodically over the whole solution space. Therefore, chaotic sequence could be employed into Eq. (4.37) to ensure the ergodicity of Φ_{ij}, shown as Eq. (4.38) by applying Eq. (4.5):

$$\Phi_{ij}^{(chaotic)} = 2 \times \left[4\Phi_{ij}(1 - \Phi_{ij})\right] - 1. \tag{4.38}$$

Step 4: Determine the abandoned food source. If a solution cannot be improved through a predetermined threshold (limited iterations), then the food source is considered to be abandoned. The employed bee will reinstate to be a scout, according to Eq. (4.39), to look for another new food source to replace the abandoned source:

$$x_{ij} = \min_j + \varphi_{ij}\left(\max_j - \min_j\right), \tag{4.39}$$

where $\max_j$ is the maximal solution, that is, $\max_j = \max\{x_{1j}, x_{2j}, \ldots, x_{Nj}\}$,$\min_j$ represents the minimal solution; thus, $\min_j = \min\{x_{1j}, x_{2j}, \ldots, x_{Nj}\}$, φ_{ij} is a random number in the range $[-1,1]$.

Similar to the parameter Φ_{ij}, φ_{ij} is also the key factor that affects convergence in ABC [50]. Thus, chaotic sequence of φ_{ij} could be employed into Eq. (4.39) to helpfully improve the ergodicity, shown as Eq. (4.40) by applying Eq. (4.5):

$$\varphi_{ij}^{(chaotic)} = 4\varphi_{ij}(1 - \varphi_{ij}). \tag{4.40}$$

Step 5: Stop criterion. If the new food source value is with smaller MAPE value or maximum iteration is reached, then the new three parameters $x_i^{(n+1)}$ and its corresponding objective value are the final solution; otherwise, go to the next iteration and go back to step 2.

4.8.3 CABC Algorithm for Three Parameters Determination and Forecasting Results

This subsection will demonstrate the three parameters determination of the proposed hybrid model (SVR with CABC), namely, SVRCABC model.

Similarly, in the training stage, the rolling-based forecasting procedure is also employed to obtain the forecasting load and receive training error in the training stage. Then, if training errors improvement occurs, the three kernel parameters—σ, C, and ε—of the SVRCABC model adjusted by CABC algorithm are employed to calculate the validation error. The adjusted parameters with minimum validation error are also selected as the most appropriate parameters. The forecasting results and the suitable parameters for the SVRCABC model are illustrated in Table 4.27, in

Table 4.27 Parameters determination of SVRCABC model

Nos. of fed-in data	Parameters			MAPE of testing (%)
	σ	C	ε	
5	668.37	117.3	4.9955	3.533
10	187.20	39.458	7.2078	3.248
15	6.73	9,058.90	12.1820	3.196
20	39.03	6,622.50	18.3850	3.176
25	**656.80**	**9,744.80**	**14.8840**	**3.164**

Table 4.28 Forecasting results of ARIMA, GRNN, TF-ε-SVR-SA, SVRABC, and SVRCABC models (unit: hundred million kWh)

Time point (month)	Actual	ARIMA (1,1,1)	GRNN ($\sigma = 3.33$)	TF-ε-SVR-SA	SVRABC	SVRCABC
Oct. 2008	181.07	192.932	191.131	184.504	184.498	182.131
Nov. 2008	180.56	191.127	187.827	190.361	183.372	182.788
Dec. 2008	189.03	189.916	184.999	202.980	183.323	182.791
Jan. 2009	182.07	191.995	185.613	195.753	183.549	182.793
Feb. 2009	167.35	189.940	184.397	167.580	183.774	182.795
Mar. 2009	189.30	183.988	178.988	185.936	183.999	182.747
Apr. 2009	175.84	189.348	181.395	180.165	183.420	182.772
MAPE (%)		6.044	4.636	3.799	3.458	3.164

Table 4.29 Wilcoxon signed-rank test

	Wilcoxon signed-rank test	
	$\alpha = 0.025$	$\alpha = 0.05$
Compared models	$W = 2$	$W = 3$
SVRCABC vs. ARIMA(1,1,1)	0[a]	0[a]
SVRCABC vs. GRNN($\sigma = 3.33$)	2[a]	2[a]
SVRCABC vs. TF-ε-SVR-SA	2[a]	2[a]
SVRCABC vs. SVRABC	0[a]	0[a]

[a]Denote that SVRCABC model significantly outperforms other alternative models

which it is also indicated that these two models all perform the best when 25 fed-in data are used.

Table 4.28 shows the actual values and the forecast values obtained using various forecasting models: ARIMA(1,1,1), GRNN($\sigma = 3.33$), TF-ε-SVR-SA, SVRABC, and SVRCABC. The MAPE values are calculated to compare fairly the proposed models with other alternative models. The proposed SVRCABC model has smaller MAPE values than ARIMA, GRNN, TF-ε-SVR-SA, and SVRABC models. Furthermore, to verify the significance of accuracy improvement of SVRCABC model comparing with ARIMA(1,1,1), GRNN($\sigma = 3.33$), TF-ε-SVR-SA, and SVRABC models, the Wilcoxon signed-rank test and asymptotic test, as mentioned, are also conducted. The test results are shown in Tables 4.29 and 4.30, respectively. Clearly, the SVRCABC model is significantly superior to

Table 4.30 Asymptotic test

Compared models	Asymptotic (S_1) test $\alpha = 0.05$	$\alpha = 0.10$
SVRCABC vs. ARIMA(1,1,1)	H_0: $e_1 = e_2$ $S_1 = -12.695$; $p = 0.000$ (reject H_0)	H_0: $e_1 = e_2$ $S_1 = -12.695$; $p = 0.000$ (reject H_0)
SVRCABC vs. GRNN($\sigma = 3.33$)	H_0: $e_1 = e_2$ $S_1 = -2.508$; $p = 0.0061$ (reject H_0)	H_0: $e_1 = e_2$ $S_1 = -2.508$; $p = 0.0061$ (reject H_0)
SVRCABC vs. TF-ε-SVR-SA	H_0: $e_1 = e_2$ $S_1 = -3.380$; $p = 0.00036$ (reject H_0)	H_0: $e_1 = e_2$ $S_1 = -3.380$; $p = 0.00036$ (reject H_0)
SVRCABC vs. SVRABC	H_0: $e_1 = e_2$ $S_1 = -9.519$; $p = 0.000$ (reject H_0)	H_0: $e_1 = e_2$ $S_1 = -9.519$; $p = 0.000$ (reject H_0)

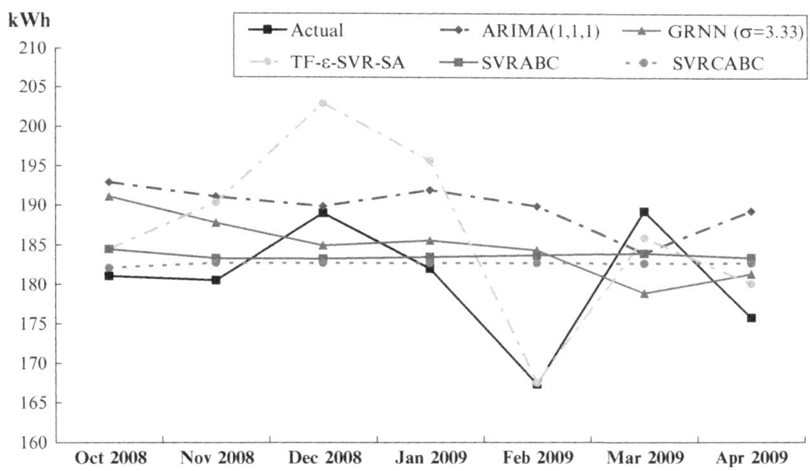

Fig. 4.17 Forecasting results of ARIMA, GRNN, TF-ε-SVR-SA, SVRABC, and SVRCABC models

ARIMA(1,1,1), GRNN($\sigma = 3.33$), TF-ε-SVR-SA, and SVRABC models. Figure 4.17 is provided to illustrate the forecasting accuracy among different models.

The superior performance of the SVRCABC model is not only caused from several similar reasons of SVRABC model, such as SVR-based model with nonlinear mapping capabilities, minimizing the structural risks rather than the training errors and the quadratic programming technique which is based on the assumptions of convex set and existence of global optimum solution, but also is caused from that the CABC algorithm employs both global search and local search in each iteration to receive better performance and applies ergodicity property of chaotic sequences to enrich the searching behavior to avoid premature convergence. Therefore, it will be theoretically approximated to the global optimum solution if the superior searching algorithm is used. For example, based on Tables 4.29 and 3.25, the CABC algorithm is excellently to shift the local solution of SVRABC model, (σ, C, ε) = (38.35, 4552.10, 16.8450) with local optimal forecasting errors, in terms of MAPE (3.458), to another better solution, (σ, C, ε) = (656.80, 9744.80, 14.8840) of SVRCABC model to be the appropriate optimal forecasting error in terms of MAPE (3.164).

Thus, it reveals that chaotic sequence is capable to avoid trapping in local minimum to achieve forecasting accuracy improvement by integrating into the CABC algorithm. It is interesting to address the SVRCABC model focuses on the interactions among individual artificial bee (dancing area and waggle dance) and well communications (waggle dance with chaotic behavior, Φ_{ij} and φ_{ij}) to negotiate and to coordinate to look for much better solutions. Therefore, the better solution evolved with "communicating in searching" activities among bees (local search) and their colony (global search) to approximately reach the global optimum or near-optimum. These two chaotic variable, Φ_{ij} and φ_{ij}, sometimes like a "broadcasting center" to exhaust their obligation to guide bees and the colony enhance its searching directions (food sources). The so-called broadcasting center is so critical for bees and the colony that, in practical, human will be stung if he unknowingly breaks the communication among bees. Finally, however, it is also clear that SVRCABC model is also not fitting the actual electric loads very well but better than other alternative models, even it has significant smaller MAPE values than other alternatives. Therefore, it also still requires hybridizing other novel mechanism to improve this shortcoming.

4.9 Chaotic Immune Algorithm in SVR's Parameters Determination

4.9.1 Shortcomings of IA and Improved by Chaotic Sequence

As mentioned in Chap. 3, Sect. 3.9, immune algorithm (IA), proposed by Mori et al. [51], is based on the learning mechanism of natural immune systems. Similar to GA, SA, and PSO algorithm, IA is also a population-based evolutionary algorithm; therefore, it provides a set of solutions for exploration and exploitation of search space to obtain optimal/near optimal solutions [52]. However, the same as GA, SA, and PSO algorithm, the diversity of the employed population set will determine the search results. Based on the IA operation procedure, if the population diversity of an initial population cannot be maintained under selective pressure, that is, the initial individuals are not necessarily fully diversified in the search space, then the IA could only seek for the solutions in the narrow space and the solution is far from the global optimum, that is, trapping in a local minimum. To overcome this drawback, it is necessary to find some effective approach and improvements of the IA to maintain the population diversity, to track in the solution space effectively and efficiently, and to avoid leading to a local optimum. One possible approach is to divide the chromosome population into several subgroups and limit the crossover between the members in different subgroups to maintain the population diversity. Such a method would require a huge population size, which is not typical in business forecasting application problem solving. Another feasible approach is focused on the chaos approach, due to its easy implementation and special ability to avoid being trapped in local optima [1].

The application of chaotic sequences can be a good alternative to diversify the initial definition domain in stochastic optimization procedures, that is, small changes in the parameter settings or the initial values in the model. Due to the ergodicity property of chaotic sequences, it will lead to very different future solution-finding behaviors; thus, chaotic sequences can be used to enrich the search behavior and to avoid being trapped in a local optimum [49]. There are lots of applications in optimization problem using chaotic sequences [7, 47, 53–55]. Coelho and Mariani [56] recently apply a chaotic artificial immune network (chaotic opt-aiNET) to solve the economic dispatch problem (EDP), based on Zaslavsky's map by its spread-spectrum characteristic and large Lyapunov exponent to successfully escape from local optimum and to converge to a stable equilibrium. Therefore, it is believable that applying chaotic sequences to diversify the initial definition domain in IA's initialization procedure [chaotic immune algorithm (CIA)] is a feasible approach to optimize the parameter selection in an SVR model. Recently, Wang et al. [57] also employed similar applications of CIA to determine the three parameters of an SVR model and obtained good performance in jumping out of the local optimum.

4.9.2 Operation Procedure of CIA

To design the CIA, many principal factors like identifying the affinity, selection of antibodies, crossover, and mutation of antibody population are similar to the IA factors, and more procedural details about the CIA used in this study are as follows, and the corresponding flowchart is shown in Fig. 4.18.

Step 1: Initialization of antibody population. The values of the three parameters in an SVR model in the *i*th iteration can be represented as $X_k^{(i)}$, $k = \sigma, C, \varepsilon$. Set $i = 0$, and we employ Eq. (4.41) to map the three parameters among the intervals (Min_k,Max_k) into chaotic variable $x_k^{(i)}$ located in the interval (0,1):

$$x_k^{(i)} = \frac{X_k^{(i)} - \mathrm{Min}_k}{\mathrm{Max}_k - \mathrm{Min}_k}. \tag{4.41}$$

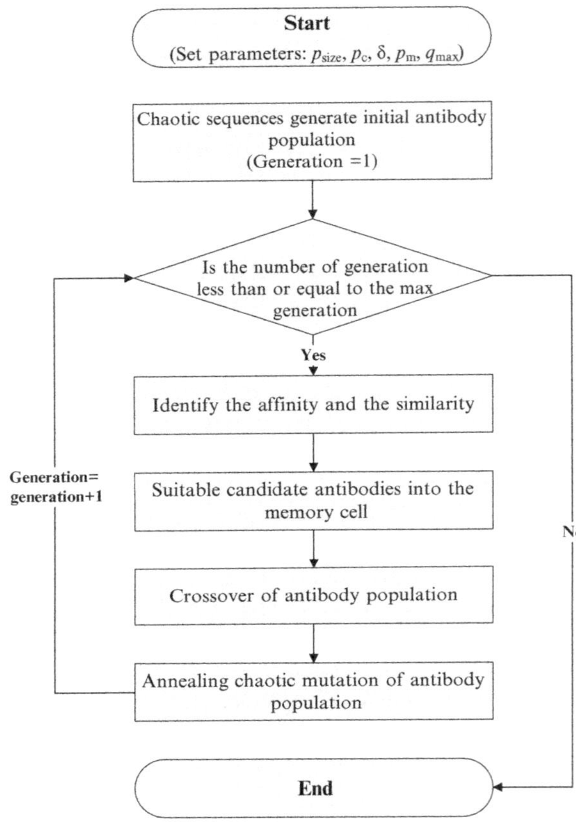

Fig. 4.18 The architecture of chaotic immune algorithm (CIA)

Then, employ the chaotic sequence, defined as Eq. (4.5) with $\mu = 4$, to compute the next iteration chaotic variable, ${x_k}^{(i+1)}$. And, transform ${x_k}^{(i+1)}$ to obtain three parameters for the next iteration, ${X_k}^{(i+1)}$ with Eq. (4.42):

$$X_k^{(i+1)} = \text{Min}_k + x_k^{(i+1)}(\text{Max}_k - \text{Min}_k). \tag{4.42}$$

After this transformation, the three parameters, σ, C, and ε, constitute the initial antibody population and then will be represented by a binary-code string. For example, assume that an antibody contains 12 binary codes to represent 3 SVR parameters. Each parameter is thus expressed by four binary codes. Assume the set boundaries for parameters σ, C, and ε are 2, 10, and 0.5, respectively; then, the antibody with binary-code "1 0 0 1 0 1 0 1 0 0 1 1" implies that the real values of the three parameters σ, C, and ε are 1.125, 3.125, and 0.09375, respectively. The number of initial antibodies is the same as the size of the memory cell. The size of the memory cell is set to 10 in this section.

Step 2: Identify the affinity and the similarity. A higher affinity value implies that an antibody has a higher activation with an antigen. To continue keeping the diversity of the antibodies stored in the memory cells, the antibodies with lower similarity have higher probability of being included in the memory cell. Therefore, an antibody with a higher affinity value and a lower similarity value has a good likelihood of entering the memory cells. The affinity between the antibody and antigen is defined as Eq. (4.43):

$$\mathrm{Ag}_k = 1/(1 + d_k), \tag{4.43}$$

where d_k denotes the SVR forecasting errors obtained by the antibody k. The similarity between antibodies is expressed as in Eq. (4.44):

$$\mathrm{Ab}_{ij} = 1/(1 + T_{ij}), \tag{4.44}$$

where T_{ij} denotes the difference between the two SVR forecasting errors obtained by the antibodies inside (existed) and outside (will be entering) the memory cell.

Step 3: Selection of antibodies in the memory cell. Antibodies with higher values of Ag_k are considered to be potential candidates for entering the memory cell. However, the potential antibody candidates with Ab_{ij} values exceeding a certain threshold are not qualified to enter the memory cell. In this investigation, the threshold value is set to 0.9.

Step 4: Crossover of antibody population. New antibodies are created via crossover and mutation operations. To perform crossover operation, strings representing antibodies are paired randomly. Moreover, the proposed scheme adopts the single-point-crossover principle. Segments of paired strings (antibodies) between two determined breakpoints are swapped. In this investigation, the probability of crossover (p_c) is set as 0.5. Finally, the three crossover parameters are decoded into a decimal format.

Step 5: Annealing chaotic mutation of antibody population. For the ith iteration (generation) crossover antibody population ($\hat{X}_k^{(i)}$, $k = \sigma, C, \varepsilon$) of current solution space (Min_k,Max_k) are mapped to chaotic variable interval [0,1] to form the crossover chaotic variable space $\hat{x}_k^{(i)}$, $k = \sigma, C, \varepsilon$, as Eq. (4.45),

$$\hat{x}_k^{(i)} = \frac{\hat{X}_k^{(i)} - \mathrm{Min}_k}{\mathrm{Max}_k - \mathrm{Min}_k}, \quad k = C,\ \sigma,\ \varepsilon,\ i = 1, 2, \ldots, q_{\max}, \tag{4.45}$$

where $q_{\max}$ is the maximum evolutional generation of the population. Then, the ith chaotic variable $x_k^{(i)}$ is summed up to $\hat{x}_k^{(i)}$, and the chaotic mutation variable is also mapped to interval [0,1] as in Eq. (4.46):

$$\tilde{x}_k^{(i)} = \hat{x}_k^{(i)} + \delta x_k^{(i)}, \tag{4.46}$$

Table 4.31 CIA's parameters setting

Population size (p_{size})	Maximal generation (q_{max})	Probability of crossover (p_c)	The annealing operation parameter (δ)	Probability of mutation (p_m)
200	500	0.5	0.9	0.1

Table 4.32 Parameters determination of SVRCIA model

Nos. of fed-in data	Parameters			MAPE of testing (%)
	σ	C	ε	
5	14.74	347.33	1.8570	4.1953
10	9.95	90.24	0.1459	3.638
15	109.06	7,298.30	11.9530	3.897
20	48.03	8,399.70	14.3720	3.514
25	**30.26**	**4,767.30**	**22.1140**	**3.0411**

where δ is the annealing operation. Finally, the chaotic mutation variable obtained in interval [0,1] is mapped to the solution interval (Min_k,Max_k) by definite probability of mutation (p_m), thus, completing a mutative operation, as shown in Eq. (4.47),

$$\tilde{X}_k^{(i)} = \text{Min}_k + \tilde{x}_k^{(i)}(\text{Max}_k - \text{Min}_k). \tag{4.47}$$

Step 6: Stopping criteria. If the number of generations equals a given scale, then the best antibody is a solution; otherwise, return to step 2.

The CIA is used to seek a better combination of the three parameters in the SVR. The value of the mean absolute percentage error (MAPE) is used as the criterion (the smallest value of MAPE) of forecasting errors to determine the suitable parameters used in SVR model in this investigation, which is given by Eq. (3.13).

4.9.3 CIA for Three Parameters Determination and Forecasting Results

This subsection will demonstrate the three parameters determination of the proposed hybrid model (SVR with CIA), namely, SVRCIA model.

The parameters of the CIA in the proposed model are experimentally set as shown in Table 4.31. Similarly, in the training stage, the rolling-based forecasting procedure is also employed to obtain the forecasting load and receive training error in the training stage. Then, if training errors improvement occurs, the three kernel parameters—σ, C, and ε—of the SVRCIA model adjusted by CIA are employed to calculate the validation error. The adjusted parameters with minimum validation error are also selected as the most appropriate parameters. The forecasting results and the suitable parameters for the SVRCIA model are illustrated in Table 4.32, in

Table 4.33 Forecasting results of ARIMA, GRNN, TF-ε-SVR-SA, SVRIA, and SVRCIA models (unit: hundred million kWh)

Time point (month)	Actual	ARIMA(1,1,1)	GRNN($\sigma = 3.33$)	TF-ε-SVR-SA	SVRIA	SVRCIA
Oct. 2008	181.07	192.932	191.131	184.504	181.322	179.0276
Nov. 2008	180.56	191.127	187.827	190.361	181.669	179.4118
Dec. 2008	189.03	189.916	184.999	202.980	183.430	179.7946
Jan. 2009	182.07	191.995	185.613	195.753	183.964	180.1759
Feb. 2009	167.35	189.940	184.397	167.580	184.030	180.5557
Mar. 2009	189.30	183.988	178.988	185.936	182.829	180.9341
Apr. 2009	175.84	189.348	181.395	180.165	183.463	178.1036
MAPE (%)		6.044	4.636	3.799	3.211	3.041

Table 4.34 Wilcoxon signed-rank test

	Wilcoxon signed-rank test	
	$\alpha = 0.025$	$\alpha = 0.05$
Compared models	$W = 2$	$W = 3$
SVRCIA vs. ARIMA(1,1,1)	0[a]	0[a]
SVRCIA vs. GRNN($\sigma = 3.33$)	1[a]	1[a]
SVRCIA vs. TF-ε-SVR-SA	1[a]	1[a]
SVRCIA vs. SVRIA	0[a]	0[a]

[a]Denote that SVRCIA model significantly outperforms other alternative models

which it is also indicated that these two models all perform the best when 25 fed-in data are used.

Table 4.33 shows the actual values and the forecast values obtained using various forecasting models: ARIMA(1,1,1), GRNN($\sigma = 3.33$), TF-ε-SVR-SA, SVRIA, and SVRCIA. The MAPE values are calculated to compare fairly the proposed models with other alternative models. The proposed SVRCIA model has smaller MAPE values than ARIMA, GRNN, TF-ε-SVR-SA, and SVRIA models. Furthermore, to verify the significance of accuracy improvement of SVRCIA model comparing with ARIMA(1,1,1), GRNN($\sigma = 3.33$), TF-ε-SVR-SA, and SVRIA models, the Wilcoxon signed-rank test and asymptotic test, as mentioned, are also conducted. The test results are shown in Tables 4.34 and 4.35, respectively. Clearly, the SVRCIA model is significantly superior to ARIMA (1,1,1), GRNN($\sigma = 3.33$), TF-ε-SVR-SA, and SVRIA models. Figure 4.19 is provided to illustrate the forecasting accuracy among different models.

The superior performance of the SVRCIA model is not only caused from several similar reasons of SVRIA model, such as SVR-based model with nonlinear mapping capabilities, minimizing the structural risks rather than the training errors and the quadratic programming technique which is based on the assumptions of convex set and existence of global optimum solution, but also is caused from that the CIA applies ergodicity property of chaotic sequences to enrich the searching behavior to avoid premature convergence. Therefore, it will be theoretically

Table 4.35 Asymptotic test

Compared models	Asymptotic (S_1) test $\alpha = 0.05$	$\alpha = 0.10$
SVRCIA vs. ARIMA(1,1,1)	H_0: $e_1 = e_2$ $S_1 = -16.145$; $p = 0.000$ (reject H_0)	H_0: $e_1 = e_2$ $S_1 = -16.145$; $p = 0.000$ (reject H_0)
SVRCIA vs. GRNN($\sigma = 3.33$)	H_0: $e_1 = e_2$ $S_1 = -3.172$; $p = 0.00076$ (reject H_0)	H_0: $e_1 = e_2$ $S_1 = -3.172$; $p = 0.00076$ (reject H_0)
SVRCIA vs. TF-ε-SVR-SA	H_0: $e_1 = e_2$ $S_1 = -3.220$; $p = 0.00064$ (reject H_0)	H_0: $e_1 = e_2$ $S_1 = -3.220$; $p = 0.00064$ (reject H_0)
SVRCIA vs. SVRIA	H_0: $e_1 = e_2$ $S_1 = -2.695$; $p = 0.00352$ (reject H_0)	H_0: $e_1 = e_2$ $S_1 = -2.695$; $p = 0.00352$ (reject H_0)

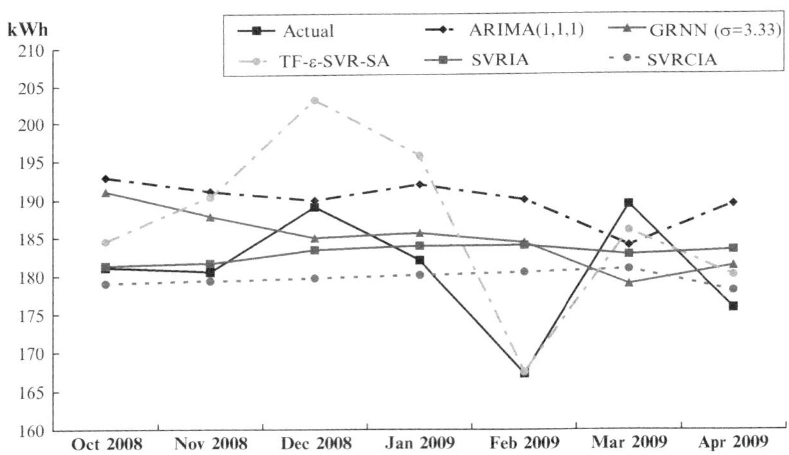

Fig. 4.19 Forecasting results of ARIMA, GRNN, TF-ε-SVR-SA, SVRIA, and SVRCIA models

approximated to the global optimum solution if the superior searching algorithm is used. For example, based on Tables 4.32 and 3.29, the CIA is excellently to shift the local solution of SVRIA model, (σ, C, ε) = (149.93, 4293.10, 9.4790) with local optimal forecasting errors, in terms of MAPE (3.211), to another better solution, (σ, C, ε) = (30.26, 4767.30, 22.1140) of SVRCIA model to be the appropriate optimal forecasting error in terms of MAPE (3.041). Thus, it reveals that chaotic sequence is capable to avoid trapping in local minimum to achieve forecasting accuracy improvement by integrating into the CIA. Finally, however, it is also clear that SVRCIA model is also not fitting the actual electric loads very well but better than other alternative models, even it has significant smaller MAPE values than other alternatives. Therefore, it also still requires hybridizing other novel mechanisms to improve this shortcoming.

References

1. Wang L, Zheng DZ, Lin QS (2001) Survey on chaotic optimization methods. Comput Technol Autom 20:1–5
2. Li D, Meng H, Shi X (1995) Membership clouds and membership cloud generators. J Comput Res Dev 32:15–20. doi:CNKI:SUN:JFYZ.0.1995-06-002
3. Li B, Jiang W (1998) Optimizing complex functions by chaos search. Cybern Syst 29:409–419. doi:10.1080/019697298125678
4. Ohya M (1998) Complexities and their applications to characterization of chaos. Int J Theor Phys 37:495–505. doi:10.1023/A:1026620313483
5. Kim JH, Stringer J (1992) Applied chaos. Wiley, New York
6. Lorenz EN (1963) Deterministic nonperiodic flow. J Atmos Sci 20:130–141. doi:10.1175/1520-0469(1963) 020<0130:DNF>2.0.CO;2
7. Liu B, Wang L, Jin YH, Tang F, Huang DX (2005) Improved particle swarm optimization combined with chaos. Chaos Soliton Fract 25:1261–1271. doi:10. 1016/j.chaos.2004.11.095
8. Cai J, Ma X, Li L, Peng H (2007) Chaotic particle swarm optimization for economic dispatch considering the generator constraints. Energ Convers Manag 48:645–653. doi:10.1016/j.enconman.2006.05.020
9. May RM (1976) Simple mathematical models with very complicated dynamics. Nature 261:459–467. doi:10.1038/261459a0
10. Sheng Y, Pan H, Xia L, Cai Y, Sun X (2010) Hybrid chaos particle swarm optimization algorithm and application in benzene-toluene flash vaporization. J Zhejiang Univ Technol 38:319–322
11. Dong Y, Guo H (2011) Adaptive chaos particle swarm optimization based on colony fitness variance. Appl Res Comput 28:855–859
12. Kao Y, Zahara E (2008) A hybrid genetic algorithm and particle swarm optimization for multimodal functions. Appl Soft Comput 8:849–857. doi:10. 1016/j.asoc.2007.07.002
13. Zhang F, Fan Y, Shen C, Li D (1999) Intelligent control based on membership cloud generators. Acta Aeronaut Et Astronaut 20:89–92
14. Yue X, Sun Z, Zhang Y, Li G (2001) Data mining technology in web logs based on the cloud model. Appl Res Comput 11:113–116
15. Wei X, Zhou Y, Huang H (2009) Adaptive particle swarm optimization algorithm based on cloud theory. Comput Eng Appl 45:48–50
16. Liu H, Zhou Y (2011) A cloud adaptive particle swarm optimization algorithm based on mean. Comput Eng Sci 33:97–100
17. Zhang Y, Shao S (2010) Cloud hyper mutation particle swarm optimization algorithm based on cloud model. Pattern Recognit Artif Intell 24:91–95. doi:CNKI:SUN:MSSB.0.2011-01-013
18. Di K, Li D, Li D (1999) Cloud theory and its applications in spatial data mining knowledge discovery. J Image Graph 4:930–935
19. Lv P, Lin Y, Zhang J (2009) Cloud theory-based simulated annealing algorithm and application. Eng Appl Artif Intel 22:742–749. doi:10.1016/j.engappai. 2009.03.003
20. Li D, Du Y (2005) Artificial intelligence with uncertainty. Chapman & Hall, Boca Raton, FL
21. Holland J (1975) Adaptation in natural and artificial system. University of Michigan Press, Ann Arbor, MI
22. Yuan X, Yuan Y, Zhang Y (2002) A Hybrid chaotic genetic algorithm for short-term hydro system scheduling. Math Comput Simul 59:319–327. doi:10. 1016/S0378-4754(01)00363-9
23. Liao GC (2006) Hybrid chaos search genetic algorithm and meta-heuristics method for short-term load forecasting. Electr Eng 88:265–276. doi:10.1007/ s00202-004-0272-0
24. Lü QZ, Shen GL, Yu RQ (2003) A chaotic approach to maintain the population diversity of genetic algorithm in network training. Comput Biol Chem 27:363–371. doi:10.1016/S1476-9271(02)00083-X

25. Yan X, Chen D, Hu S (2003) Chaos-genetic algorithms for optimizing the operating conditions based on RBF-PLS model. Comput Chem Eng 27:1393–1404. doi:10.1016/S0098-1354(03) 00074-7
26. Cheng CT, Wang WC, Xu DM, Chau KW (2008) Optimizing hydropower reservoir operation using hybrid genetic algorithm and chaos. Water Resour Manag 22:895–909. doi:10.1007/ s11269-007-9200-1
27. Chen L, Aihara K (1995) Chaotic simulated annealing by a neural network model with transient chaos. Neural Netw 8:915–930. doi:10.1016/0893-6080(95) 00033-V
28. Wang L, Smith K (1998) On chaotic simulated annealing. IEEE Trans Neural Netw 9:716–718. doi:10.1109/72.701185
29. Alizadeh S, Ghazanfari M (2009) Learning FCM by chaotic simulated annealing. Chaos Solitons Fractals 41:1182–1190. doi:10.1016/j.chaos.2008.04.058
30. Metropolis N, Rosenbluth AW, Rosenbluth MN, Teller AH (1953) Equations of state calculations by fast computing machines. J Chem Phys 21:1087–1092. doi:10.1063/1.1699114
31. Kirkpatrick S, Gelatt CD, Vecchi MP (1983) Optimization by simulated annealing. Science 220:671–680. doi:10.2307/1690046
32. Van Laarhoven PJM, Aarts EHL (1987) Simulated annealing: theory and applications. Kluwer, Dordrecht
33. Chen G, Mao Y, Chui CK (2004) A symmetric image encryption scheme based on 3D chaotic cat maps. Chaos Solitons Fractals 21:749–761. doi:10.1016/j.chaos. 2003.12.022
34. Li D, Cheung D, Shi XM, Ng V (1998) Uncertainty reasoning based on cloud models in controllers. Comput Math Appl 35:99–123. doi:10.1016/S0898-1221 (97)00282-4
35. Wang S, Li D, Shi W, Li D, Wang Z (2003) Cloud model-based spatial data mining. Geogr Info Sci 9(2):60–70. doi:10.1080/10824000309480589
36. Wang H, Deng Y (2007) Spatial clustering method based on cloud model. In: Proceedings of the fourth international conference on fuzzy systems and knowledge discovery 7:272–276. doi:10.1109/FSKD.2007.526
37. Zhu Y, Dai C, Chen W (2005) Adaptive probabilities of crossover and mutation in genetic algorithms based on cloud generators. J Comput Inf Syst 1:671–678
38. Aarts EHL, Korst JHM (1989) Simulated annealing and Boltzmann machines. Wiley, New York
39. Anderson HR, McGeehan JP (1994) Optimizing microcell base station locations using simulated annealing techniques. In: Proceedings of 44th IEEE Vehicular Technology conference 2:858–862. doi:10.1109/VETEC.1994.345212
40. Dekkers A, Aarts EHL (1991) Global optimization and simulated annealing. Math Program 50:367–393. doi:10.1007/BF01594945
41. Angeline PJ (1998) Evolutionary optimization versus particle swarm optimization: philosophy and performance differences. In: Proceedings of the 7th international conference on evolutionary programming, San Diego, CA, pp 601–610. doi:10.1007/BFb0040811
42. Kennedy J, Eberhart RC (1995) Particle swarm optimization. In: Proceeding of IEEE international conference neural networks, Washington, DC, pp 1942–1948. doi:10.1109/ ICNN.1995.488968
43. Dorigo M (1992) Optimization, learning, and natural algorithms. Doctoral dissertation, Dipartimento di Elettronica e Informazione, Politecnico di Milano, Italy
44. Dorigo M, Maniezzo V, Colorni A (1996) Ant system: optimization by a colony of cooperating ants. IEEE Trans Syst Man Cybern B Cybern 26:29–41. doi:10.1109/3477.484436
45. Li Y, Wen Q, Li L, Peng H (2009) Hybrid chaotic ant swarm optimization. Chaos Solitons Fractals 42:880–889. doi:10.1016/j.chaos.2009.02.020
46. Cole BJ (1991) Is animal behavior chaotic? Evidence from the activity of ants. Proc R Soc Lond B 244:253–259. doi:10.1098/rspb.1991.0079
47. Li L, Yang Y, Peng H, Wang X (2006) Parameters identification of chaotic systems via chaotic ant swarm. Chaos Solitons Fractals 28:1204–1211. doi:10.1016/j.chaos.2005.04.110

48. Sadjadi F (2004) Comparison of fitness scaling functions in genetic algorithms with applications to optical processing. Proc SPIE 5557:356–364. doi:10.1117/12.563910
49. Pan H, Wang L, Liu B (2008) Chaotic annealing with hypothesis test for function optimization in noisy environments. Chaos Solitons Fractals 35:888–894. doi:10. 1016/j.chaos.2006.05.070
50. Singh A (2009) An artificial bee colony algorithm for the leaf-constrained minimum spanning tree problem. Appl Soft Comput 9:625–631. doi:10.1016/j. asoc.2008.09.001
51. Mori K, Tsukiyama M, Fukuda T (1993) Immune algorithm with searching diversity and its application to resource allocation problem. Trans Inst Electr Eng Jpn 113-C:872–878
52. Prakash A, Khilwani N, Tiwari MK, Cohen Y (2008) Modified immune algorithm for job selection and operation allocation problem in flexible manufacturing system. Adv Eng Softw 39:219–232. doi:10.1016/j.advengsoft. 2007.01.024
53. Zuo XQ, Fan YS (2006) A chaos search immune algorithm with its application to neuro-fuzzy controller design. Chaos Solitons Fractals 30:94–109. doi:10.1016/j. chaos.2005.08.126
54. Yang D, Li G, Cheng G (2007) On the efficiency of chaos optimization algorithms for global optimization. Chaos Solitons Fractals 34:1366–1375. doi:10.1016/j.chaos.2006.04.057
55. Tavazoei MS, Haeri M (2007) Comparison of different one-dimensional maps as chaotic search pattern in chaos optimization algorithms. Appl Math Comput 187:1076–1085. doi:10.1016/j.amc.2006.09.087
56. Coelho LDS, Mariani VC (2009) Chaotic artificial immune approach applied to economic dispatch of electric energy using thermal units. Chaos Solitons Fractals 40:2376–2383. doi:10.1016/j.chaos.2007.10.032
57. Wang J, Wang Y, Zhang C, Du W, Zhou C, Liang Y (2009) Parameter selection of support vector regression based on a novel chaotic immune algorithm. In: Proceedings of the 4th international conference on innovative computing, information and control, pp 652-655. doi: 10.1109/ICICIC.2009.287

Chapter 5
Recurrent/Seasonal Mechanism to Improve the Accurate Level of Forecasting

As demonstrated in Chap. 4, different hybrid chaotic evolutionary algorithms, including chaotic genetic algorithm (CGA), chaotic simulated annealing (CSA) algorithm, chaotic cloud simulated annealing (CCSA) algorithm, chaotic GASA (CGASA) algorithm, chaotic particle swarm optimization (CPSO) algorithm, chaotic ant swarm (CAS) optimization algorithm, chaotic artificial bee colony (CABC) algorithm, and chaotic immune algorithm (CIA), are employed to determine suitable parameter combination of an SVR-based electric load forecasting model. These forecasting results indicate that almost all SVR-based models with different hybrid chaotic evolutionary algorithms are superior to other competitive forecasting models (including ARIMA, GRNN, and TF-ε-SVR-SA models). However, these hybrid chaotic evolutionary algorithms still do not provide satisfactory forecasting performance (well fitting the actual fluctuation tendency) even their forecasting accuracy receives significant level. To improve the fitting effects for each SVR-chaotic-/cloud-evolutionary algorithm-based model, this chapter introduces two combined mechanisms (recurrent mechanism or seasonal mechanism) to significantly improve the fitting effects with the actual fluctuation tendency.

5.1 Combined Mechanisms

5.1.1 Recurrent Mechanism

For a feed-forward neural network, links may be established within layers of a neural network; these types of networks are so-called recurrent neural networks (RNNs). The main concept on which RNNs are based is that every unit is considered as an output of the network and the provision of adjusted information as input in a training process [1]. RNNs are extensively applied in time series forecasting, such as Jordan recurrent neural network model [2] (Fig. 1.1), Elman recurrent neural network model [3] (Fig. 1.2), and Williams and Zipser recurrent neural network model [4] (Fig. 1.3). These three models mentioned all consist of multilayer perceptron (MLP) with a

W.-C. Hong, *Intelligent Energy Demand Forecasting*, Lecture Notes in Energy 10, DOI 10.1007/978-1-4471-4968-2_5,

hidden layer. Jordan networks have a feedback loop from the output layer with past values to an additional input, namely, "context layer." Then, output values from the context layer are fed back into the hidden layer. Elman networks have a feedback loop from the hidden layer to the context layer. In Williams and Zipser networks, nodes in the hidden layer are fully connected to each other. Both Jordan and Elman networks include an additional information source from the output layer or the hidden layer. Hence, these models use mainly past information to capture detailed information. Williams and Zipser network takes much more information from the hidden layer and back into themselves. Therefore, Williams and Zipser networks are sensitive when models are implemented [5]. Jordan networks and Elman networks are suited to time series forecasting [6, 7]. In this book, the Jordan network is employed as a base to construct the recurrent SVR models.

In a Jordan recurrent neural network, all neurons in a layer except those in the context layer are connected with all neurons in the next layer. A context layer is a special hidden layer. Interactions only occur between neurons in the hidden layer and those in the context layer. For a Jordan network with p inputs, q hidden, and r output neurons, the output of the nth neuron, $f_n(t)$, is shown as Eq. (5.1) [8–11]:

$$f_n(t) = \sum_{i=1}^{q} W_i \varphi_i(t) + b_i(t) \tag{5.1}$$

where W_i are weights between the hidden and output layers and $\varphi_i(t)$ is the output function of the hidden neurons, which is as Eq. (5.2),

$$\varphi_i(t) = g\left(\sum_{j=1}^{P} v_{ij} x_j(t) + \sum_{k=1}^{s} \sum_{v=1}^{r} w_{ikv} f_v(t-k) + b_i(t)\right), \tag{5.2}$$

where v_{ij} are weights between the input and the hidden layer, w_{ikv} are weights between the context and the hidden layer with k delay periods, and s is the total number of context layers in past output data.

Back-propagation yields gradients for adapting weights of a neural network. The back-propagation algorithm is presented as follows. First, the output of the nth neuron in Eq. (5.2) is rewritten as

$$f_n(t) = h(x^T(t)\phi(t)), \tag{5.3}$$

where $h(\cdot)$ is the nonlinearity function of $x^T(t)$ and $f_n(t)$; $x^T(t) = [x_1(t), \ldots, x_p(t)]^T$ is the input vector; and $\phi(t) = [\phi_1(t), \ldots, \phi_P(t)]^T$ is the weight vector. A cost function is then presented to be the instantaneous performance index, as shown in Eq. (5.4):

$$J(\phi(t)) = \frac{1}{2}[d(t) - f_n(t)]^2 = \frac{1}{2}\left[d(t) - h(x^T(t)\phi(t))\right]^2 \tag{5.4}$$

where $d(t) = [d_1(t), \ldots, d_p(t)]^T$ is the desired output.

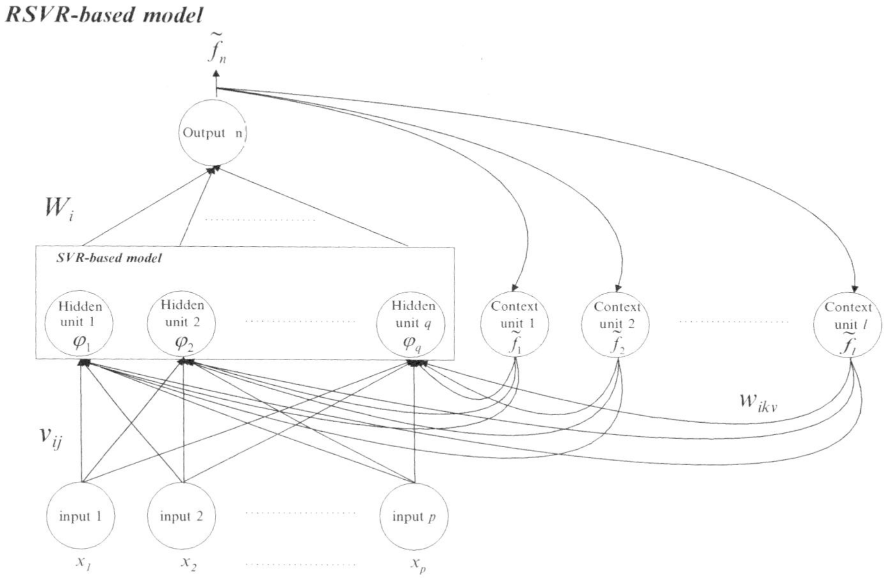

Fig. 5.1 The architecture of RSVR-based model

Second, the instantaneous output error at the output neuron and the revised weight vector in the next moment are given by Eqs. (5.5) and (5.6), respectively:

$$e(t) = d(t) - f_n(t) = d(t) - h(x^T(t)\phi(t)), \tag{5.5}$$

$$\phi(t+1) = \phi(t) - \eta \nabla_{\phi} J(\phi(t)), \tag{5.6}$$

where η is the learning rate.

Third, the gradient $\nabla_{\phi} J(\phi(t))$ can be calculated as Eq. (5.7):

$$\nabla_{\phi} J(\phi(t)) = \frac{\partial J \phi(t)}{\partial \phi(t)} = e(t) \times \frac{\partial e(t)}{\partial \phi(t)} = -e(t)h'(x^T(t)\phi(t))x(t), \tag{5.7}$$

where $h'(\cdot)$ is the first derivative of the nonlinearity $h(\cdot)$. Finally, the weight is revised as Eq. (5.8):

$$\phi(t+1) = \phi(t) + \eta e(t)h'(x^T(t)\phi(t))x(t) \tag{5.8}$$

Figure 5.1 shows the architecture of the general recurrent SVR-based (RSVR-based) model. The output of RSVR-based model ($\tilde{f}_n(t)$) is as Eq. (5.9):

$$\tilde{f}_n(t) = \sum_{i=1}^{P} W^T \psi(x^T(t)) + b(t) \tag{5.9}$$

Then, Eq. (5.9) replaces Eq. (2.47) in the SVR-based model, to run the loop of SVR-based model in the search for values of three parameters. Finally, the forecast values $\tilde{f}_n(t)$ are calculated using Eq. (5.9).

5.1.2 Seasonal Mechanism

As mentioned, the electric load often demonstrates a cyclic tendency due to economic activities or climate cyclic nature. Lots of researchers in financial fields have explored how to identify the seasonal index to adjust the seasonal biases, such as Martens et al. [12], Taylor and Xu [13], and Andersen and Bollerslev [14] apply flexible Fourier form to estimate the variation of daily stock exchange, then receive the seasonal variation estimator; Deo et al. [15] proposed a revised model to further identify the seasonal variation estimator that is composed of two linear combinations in a cyclic period. Based on the data series-type consideration and inspired from previous papers, this investigation firstly applies ARIMA methodology to identify the seasonal length, then proposes the seasonal index to easily adjust cyclic effects, as shown in Eq. (5.10):

$$\text{Season}_t = \ln\left(\frac{a_t}{f_t}\right)^2 = 2\left(\ln a_t - \ln \sum_{i=1}^{n} (\beta_i^* - \beta_i)K(x, x_i) + b\right) \tag{5.10}$$

where $t = j, l + j, 2l + j, \ldots, (m - 1)l + j$ only for the same time point in each period. Then, the seasonal index (SI) for each time point j is computed as Eq. (5.11):

$$\text{SI}_j = \exp\left(\frac{1}{m}(\text{season}_j + \text{season}_{l+j} + \cdots + \text{season}_{(m-1)l+j})\right)\Big/2 \tag{5.11}$$

where $j = 1, 2, \ldots, l$. The seasonal mechanism is shown in Fig. 5.2.

Eventually, the forecasting value of the SSVRCCSA is obtained by Eq. (5.12):

$$f_{N+k} = \left(\sum_{i=1}^{N} (\beta_i^* - \beta_i)K(x_i, x_{N+k}) + b\right) \times \text{SI}_k, \tag{5.12}$$

where $k = 1, 2, \ldots, l$ implies the time point in another period (for forecasting period).

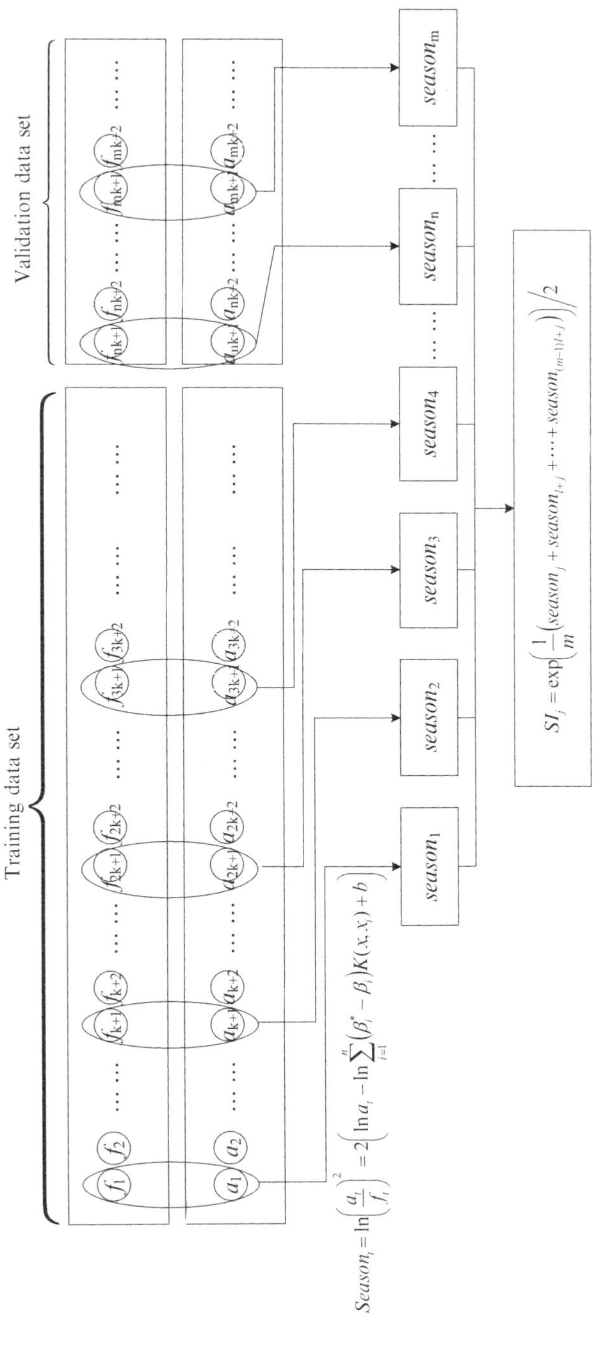

Fig. 5.2 The process of seasonal mechanism

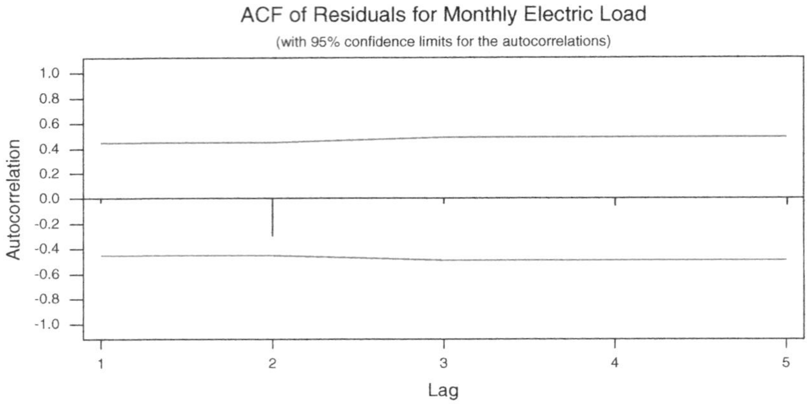

Fig. 5.3 Estimated residual ACF

5.2 Seasonal ARIMA Model and Seasonal HW (SHW) Model

5.2.1 SARIMA Model

For the Seasonal ARIMA (SARIMA) model, by Minitab 14 statistic software, the parameters are determined by taking the first-order regular difference and first seasonal difference to remove nonstationary and seasonality characteristics. Using statistical packages, with no residuals autocorrelated and approximately white noise residuals, the most suitable models for the employed electric load data is SARIMA(4,2,2) $\times$ $(1,2,1)_{12}$ with constant item. The equation used for the SARIMA model is presented as Eq. (5.13):

$$\begin{aligned}(1 + 1.067B + 0.6578B^2 + 0.4569B^3 + 0.1819B^4)(1 + 0.3012B^5)W_{\mathrm{t}} = \\ -0.7758 + (1 - 0.8055B - 0.1857B^2)(1 - 0.5054B^5)\varepsilon_{\mathrm{t}},\end{aligned} \quad (5.13)$$

where $W_t = (1 - B)^2(1 - B^{12})^2X_t$.

After determining the suitable parameters of the SARIMA model, it is important to examine how closely the proposed model fits a given time series. The autocorrelation function (ACF) is calculated to verify the parameters. Figure 5.3 plots the estimated residual ACF and indicates that the residuals are not autocorrelated. PACF, the partial autocorrelation function, displayed in Fig. 5.4, is also used to check the residuals and indicates that the residuals are not correlated. The forecasting results are shown in the third column of Table 5.1.

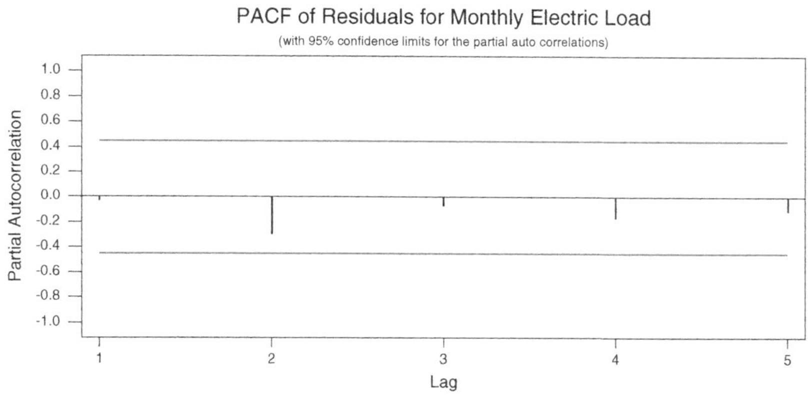

Fig. 5.4 Estimated residual PACF

Table 5.1 Forecasting results of SARIMA, SHW, GRNN, and BPNN models (unit: hundred million kWh)

Time point (month)	Actual	SARIMA(4,2,2) $\times$ $(1,2,1)_{12}$	SHW(0.12, 0.95, 0.2, 0.2)	GRNN ($\sigma = 3.33$)	BPNN
Oct. 2008	181.07	184.210	181.554	191.131	172.084
Nov. 2008	180.56	187.638	190.312	187.827	172.597
Dec. 2008	189.03	194.915	197.887	184.999	176.614
Jan. 2009	182.07	197.119	193.511	185.613	177.641
Feb. 2009	167.35	155.205	163.113	184.397	180.343
Mar. 2009	189.30	187.090	181.573	178.988	183.830
Apr. 2009	175.84	166.394	178.848	181.395	187.104
MAPE (%)		4.404	3.566	4.636	5.062

5.2.2 *SHW Model*

For the seasonal Holt–Winters (SHW) model, by Minitab 14 statistic software, the α-value and β-value are determined as 0.5618 and 0.0472, respectively.

For the seasonal Holt–Winters (SHW) method, by Minitab 14 statistic software, the appropriate parameters (L, α, β, and γ) are determined 12, 0.95, 0.20, and 0.20, correspondingly. The forecasting results are shown in the fourth column of Table 5.1.

Figure 5.5 is provided to illustrate the forecasting accuracy among different models. Obviously, these four models, excepting GRNN and BPNN models, are fitting much better than ARIMA and HW models. Furthermore, to verify the significance of accuracy improvement of SARIMA(4,2,2) $\times$ $(1,2,1)_{12}$ and SHW (0.12, 0.95, 0.2, 0.2) models comparing with ARIMA(1,1,1) and HW(0.5618, 0.0472) models, respectively, the Wilcoxon signed-rank test and asymptotic test

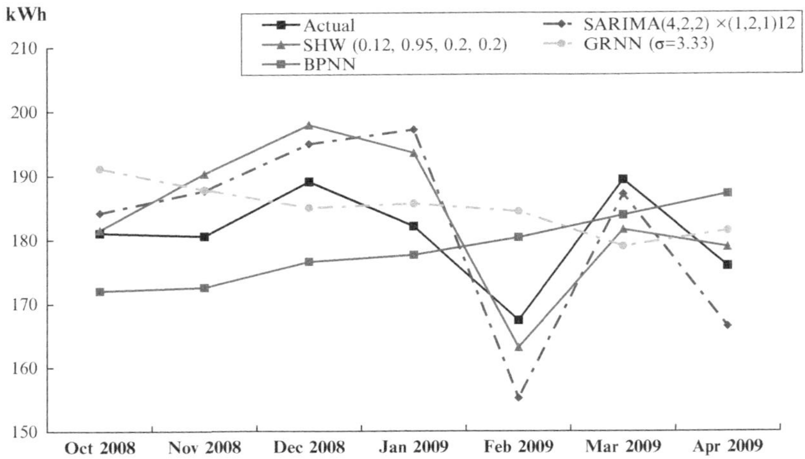

Fig. 5.5 Forecasting results of SARIMA, SHW, GRNN, and BPNN models

Table 5.2 Wilcoxon signed-rank test

	Wilcoxon signed-rank test	
	$\alpha = 0.025$	$\alpha = 0.05$
Compared models	$W = 2$	$W = 3$
SARIMA(4,2,2) × $(1,2,1)_{12}$ vs. ARIMA(1,1,1)	2[a]	2[a]
SHW(0.12, 0.95, 0.2, 0.2) vs. HW(0.5618, 0.0472)	1[a]	1[a]

[a]Denotes that SARIMA and SHW models significantly outperform other alternative models

Table 5.3 Asymptotic test

	Asymptotic (S_1) test	
Compared models	$\alpha = 0.05$	$\alpha = 0.10$
SARIMA(4,2,2) × $(1,2,1)_{12}$ vs. ARIMA(1,1,1)	H_0: $e_1 = e_2$ $S_1 = -9.511$; $p = 0.000$ (reject H_0)	H_0: $e_1 = e_2$ $S_1 = -5.958$; $p = 0.000$ (reject H_0)
SHW(0.12, 0.95, 0.2, 0.2) vs. HW (0.5618, 0.0472)	H_0: $e_1 = e_2$ $S_1 = -6.262$; $p = 0.000$ (reject H_0)	H_0: $e_1 = e_2$ $S_1 = -6.262$; $p = 0.000$ (reject H_0)

are conducted and shown in Tables 5.2 and 5.3. It is clear to receive that SARIMA and SHW models significantly outperform ARIMA and HW models, respectively. Therefore, SARIMA(4,2,2) × $(1,2,1)_{12}$ and SHW(0.12, 0.95, 0.2, 0.2) are potential to compare with seasonal-SVR-chaotic-evolutionary algorithm-based models in the following sections.

Table 5.4 The seasonal indexes for each time point (month) for the SVRCGA model

Time point (month)	Seasonal index	Time point (month)	Seasonal index
January	1.0163	July	1.0566
February	0.9057	August	1.0527
March	1.0085	September	0.9987
April	0.9834	October	0.9726
May	1.0112	November	1.0237
June	1.0140	December	1.0627

Table 5.5 Forecasting results of SARIMA, SHW, TF-ε-SVR-SA, SVRCGA, and SSVRCGA models (unit: hundred million kWh)

Time point (month)	Actual	SARIMA $(4,2,2) \times (1,2,1)_{12}$	SHW(0.12, 0.95, 0.2, 0.2)	TF-ε-SVR-SA	SVRCGA	SSVRCGA
Oct. 2008	181.07	184.210	181.554	184.504	185.224	180.1534
Nov. 2008	180.56	187.638	190.312	190.361	186.046	190.4631
Dec. 2008	189.03	194.915	197.887	202.980	186.865	198.5843
Jan. 2009	182.07	197.119	193.511	195.753	187.680	190.7387
Feb. 2009	167.35	155.205	163.113	167.580	188.493	170.7151
Mar. 2009	189.30	187.090	181.573	185.936	189.149	190.7486
Apr. 2009	175.84	166.394	178.848	180.165	178.300	175.3391
MAPE (%)		4.404	3.566	3.799	3.382	2.695

5.3 Seasonal Mechanism in SVRCGA Model and Forecasting Results

Based on the total employed electric load, each fixed point (month) has its electric load status (specific data pattern); therefore, the seasonal (cyclic) length can be estimated as 12 [16]. The 12 seasonal indexes can be estimated by the 46 in-sample forecasting loads of the SVRCGA model mentioned in Sect. 4.2.3, including 32 and 14 in-sample forecasting loads in training stage and validation stage, respectively, as shown in Table 5.4. The actual values and the out-of-sample forecasting loads obtained by different forecasting models, including SARIMA$(4,2,2) \times (1,2,1)_{12}$, TF-ε-SVR-SA, SHW(0.12, 0.95, 0.2, 0.2), SVRCGA, and SSVRCGA models, are illustrated in Table 5.5. The proposed SSVRCGA model with smaller MAPE values is superior to SARIMA$(4,2,2) \times (1,2,1)_{12}$, SHW(0.12, 0.95, 0.2, 0.2), TF-ε-SVR-SA, and SVRCGA models, due to its capability to excellently learn about the monthly load changing tendency. The seasonal mechanism further revises the forecasting results from the SVRCGA model (MAPE = 3.382 %), based on the seasonal indexes (per month) obtained from training and validation stages, to achieve more acceptable forecasting accuracy (2.695 %).

Furthermore, for forecasting accuracy improvement significant test, the Wilcoxon signed-rank test and asymptotic test, as mentioned, are also conducted. The test results are shown in Tables 5.6 and 5.7, respectively. Clearly, the SSVRCGA model receives complete significant forecasting accuracy improvement than SARIMA$(4,2,2) \times (1,2,1)_{12}$ model, but incomplete significant forecasting accuracy

Table 5.6 Wilcoxon signed-rank test

	Wilcoxon signed-rank test	
	$\alpha = 0.025$	$\alpha = 0.05$
Compared models	$W = 2$	$W = 3$
SSVRCGA vs. SARIMA(4,2,2) × $(1,2,1)_{12}$	2[a]	2[a]
SSVRCGA vs. SHW(0.12, 0.95, 0.2, 0.2)	3	3[a]
SSVRCGA vs. TF-ε-SVR-SA	3	3[a]
SSVRCGA vs. SVRCGA	3	3[a]

[a]Denotes that SSVRCGA model significantly outperforms other alternative models

Table 5.7 Asymptotic test

	Asymptotic (S_1) test	
Compared models	$\alpha = 0.05$	$\alpha = 0.10$
SSVRCGA vs. SARIMA (4,2,2) × $(1,2,1)_{12}$	H_0: $e_1 = e_2$ $S_1 = -2.958$; $p = 0.00155$ (reject H_0)	H_0: $e_1 = e_2$ $S_1 = -2.958$; $p = 0.00155$ (reject H_0)
SSVRCGA vs. SHW(0.12, 0.95, 0.2, 0.2)	H_0: $e_1 = e_2$ $S_1 = -3.146$; $p = 0.000828$ (reject H_0)	H_0: $e_1 = e_2$ $S_1 = -3.146$; $p = 0.000828$ (reject H_0)
SSVRCGA vs. TF-ε-SVR-SA	H_0: $e_1 = e_2$ $S_1 = -4.284$; $p = 0.000$ (reject H_0)	H_0: $e_1 = e_2$ $S_1 = -4.284$; $p = 0.000$ (reject H_0)
SSVRCGA vs. SVRCGA	H_0: $e_1 = e_2$ $S_1 = -3.180$; $p = 0.00074$ (reject H_0)	H_0: $e_1 = e_2$ $S_1 = -3.180$; $p = 0.00074$ (reject H_0)

improvement than SHW(0.12, 0.95, 0.2, 0.2), TF-ε-SVR-SA, and SVRCGA models (only receives significance with $\alpha = 0.05$ level in Wilcoxon test, and all pass with both levels in asymptotic test). Particularly for comparing with TF-ε-SVR-SA model (also with seasonal adjustment mechanism but without hybrid evolutionary algorithm and chaotic sequence), the comparison results recognize that chaotic sequence could significantly improve the performance in terms of premature convergence. By comparing SVRCGA with SSVRCGA models, it also indicates the significant superiority from seasonal mechanism, even it is a little time-consuming; however, it deserves to pay some attention on those cyclic information while modeling. Figure 5.6 is provided to illustrate the forecasting accuracy among different models.

5.4 Seasonal Mechanism in SVRCSA Model and Forecasting Results

Similarly, the seasonal (cyclic) length of the total employed electric load is also set as 12. Thus, the 12 seasonal indexes are estimated by the 46 in-sample forecasting loads of the SVRCSA model mentioned in Sect. 4.3.3, including 32 and 14 in-sample

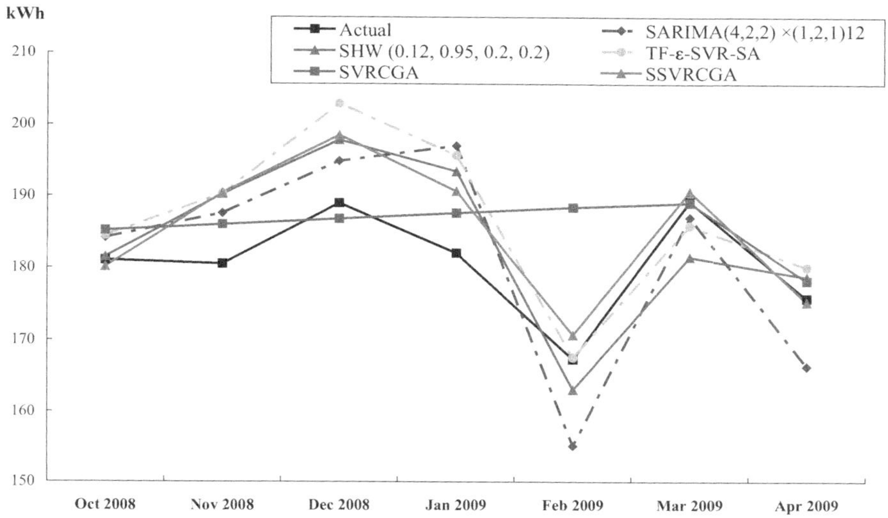

Fig. 5.6 Forecasting results of SARIMA, SHW, TF-ε-SVR-SA, SVRCGA, and SSVRCGA models

Table 5.8 The seasonal indexes for each time point (month) for the SVRCSA model

Time point (month)	Seasonal index	Time point (month)	Seasonal index
January	1.0170	July	1.0714
February	0.9212	August	1.0633
March	1.0324	September	1.0065
April	0.9988	October	0.9894
May	1.0302	November	1.0430
June	1.0301	December	1.0617

forecasting loads in training stage and validation stage, respectively, as shown in Table 5.8. The actual values and the out-of-sample forecasting loads obtained by different forecasting models, including SARIMA(4,2,2) × $(1,2,1)_{12}$, TF-ε-SVR-SA, SHW(0.12, 0.95, 0.2, 0.2), SVRCSA, and SSVRCSA models, are illustrated in Table 5.9. The proposed SSVRCSA model with smaller MAPE values is superior to SARIMA(4,2,2) × $(1,2,1)_{12}$, SHW(0.12, 0.95, 0.2, 0.2), TF-ε-SVR-SA, and SVRCSA models, due to its capability to excellently learn about the monthly load changing tendency. The seasonal mechanism further revises the forecasting results from the SVRCSA model (MAPE = 3.633 %), based on the seasonal indexes (per month) obtained from training and validation stages, to achieve more acceptable forecasting accuracy (2.844 %).

For forecasting accuracy improvement significant test, the Wilcoxon signed-rank test and asymptotic test are employed. The test results are shown in Tables 5.10 and 5.11, respectively. Clearly, the SSVRCSA model receives complete significant forecasting accuracy improvement than SARIMA(4,2,2) × $(1,2,1)_{12}$ and SHW

Table 5.9 Forecasting results of SARIMA, SHW, TF-ε-SVR-SA, SVRCSA, and SSVRCSA models (unit: hundred million kWh)

Time point (month)	Actual	SARIMA $(4,2,2) \times (1,2,1)_{12}$	SHW(0.12, 0.95, 0.2, 0.2)	TF-ε-SVR-SA	SVRCSA	SSVRCSA
Oct. 2008	181.07	184.210	181.554	184.504	184.059	182.103
Nov. 2008	180.56	187.638	190.312	190.361	183.717	191.626
Dec. 2008	189.03	194.915	197.887	202.980	183.854	195.202
Jan. 2009	182.07	197.119	193.511	195.753	184.345	187.487
Feb. 2009	167.35	155.205	163.113	167.580	184.489	169.942
Mar. 2009	189.30	187.090	181.573	185.936	184.186	190.149
Apr. 2009	175.84	166.394	178.848	180.165	184.805	184.576
MAPE (%)		4.404	3.566	3.799	3.633	2.844

Table 5.10 Wilcoxon signed-rank test

	Wilcoxon signed-rank test	
	$\alpha = 0.025$	$\alpha = 0.05$
Compared models	$W = 2$	$W = 3$
SSVRCSA vs. SARIMA$(4,2,2) \times (1,2,1)_{12}$	2[a]	2[a]
SSVRCSA vs. SHW(0.12, 0.95, 0.2, 0.2)	2[a]	2[a]
SSVRCSA vs. TF-ε-SVR-SA	3	3[a]
SSVRCSA vs. SVRCSA	3	3[a]

[a]Denotes that SSVRCSA model significantly outperforms other alternative models

Table 5.11 Asymptotic test

	Asymptotic (S_1) test	
Compared models	$\alpha = 0.05$	$\alpha = 0.10$
SSVRCSA vs. SARIMA $(4,2,2) \times (1,2,1)_{12}$	H_0: $e_1 = e_2$ $S_1 = -2.657$; $p = 0.00394$ (reject H_0)	H_0: $e_1 = e_2$ $S_1 = -2.657$; $p = 0.00394$ (reject H_0)
SSVRCSA vs. SHW(0.12, 0.95, 0.2, 0.2)	H_0: $e_1 = e_2$ $S_1 = -2.294$; $p = 0.01088$ (reject H_0)	H_0: $e_1 = e_2$ $S_1 = -2.294$; $p = 0.01088$ (reject H_0)
SSVRCSA vs. TF-ε-SVR-SA	H_0: $e_1 = e_2$ $S_1 = -3.465$; $p = 0.000265$ (reject H_0)	H_0: $e_1 = e_2$ $S_1 = -3.465$; $p = 0.000265$ (reject H_0)
SSVRCSA vs. SVRCSA	H_0: $e_1 = e_2$ $S_1 = -2.093$; $p = 0.0182$ (reject H_0)	H_0: $e_1 = e_2$ $S_1 = -2.093$; $p = 0.0182$ (reject H_0)

(0.12, 0.95, 0.2, 0.2) models, but incomplete significant forecasting accuracy improvement than TF-ε-SVR-SA and SVRCSA models (only receives significance with $\alpha = 0.05$ level in Wilcoxon test, and all pass with both levels in asymptotic test). Particularly for comparing with TF-ε-SVR-SA model (also with seasonal adjustment mechanism but without hybrid evolutionary algorithm and chaotic sequence), the

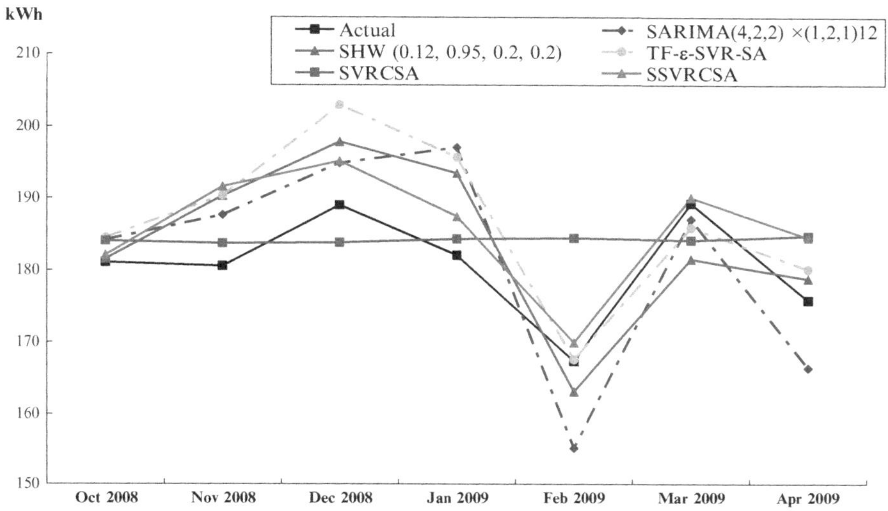

Fig. 5.7 Forecasting results of SARIMA, SHW, TF-ε-SVR-SA, SVRCSA, and SSVRCSA models

Table 5.12 The seasonal indexes for each month for the SVRCCSA model

Time point (month)	Seasonal index	Time point (month)	Seasonal index
January	1.0207	July	1.0891
February	0.9391	August	1.0789
March	1.0500	September	1.0258
April	1.0056	October	1.0053
May	1.0418	November	1.0612
June	1.0466	December	1.0643

comparison results also recognize that chaotic sequence could significantly improve the performance in terms of premature convergence. By comparing SVRCSA with SSVRCSA models, it also indicates the significant superiority from seasonal mechanism. It deserves to pay some attention on those cyclic information while modeling. Figure 5.7 is provided to illustrate the forecasting accuracy among different models.

5.5 Seasonal Mechanism in SVRCCSA Model and Forecasting Results

The seasonal (cyclic) length of the total employed electric load is also set as 12. Thus, the 12 seasonal indexes are estimated by the 46 in-sample forecasting loads of the SVRCCSA model mentioned in Sect. 4.4.3, including 32 and 14 in-sample forecasting loads in training stage and validation stage, respectively, as shown in Table 5.12. The actual values and the out-of-sample forecasting loads obtained by

Table 5.13 Forecasting results of SARIMA, SHW, TF-ε-SVR-SA, SVRCCSA, and SSVRCCSA models (unit: hundred million kWh)

Time point (month)	Actual	SARIMA $(4,2,2)\times(1,2,1)_{12}$	SHW(0.12, 0.95, 0.2, 0.2)	TF-ε-SVR-SA	SVRCCSA	SSVRCCSA
Oct. 2008	181.07	184.210	181.554	184.504	179.138	180.083
Nov. 2008	180.56	187.638	190.312	190.361	179.789	190.786
Dec. 2008	189.03	194.915	197.887	202.980	179.834	191.389
Jan. 2009	182.07	197.119	193.511	195.753	179.835	183.551
Feb. 2009	167.35	155.205	163.113	167.580	179.835	168.878
Mar. 2009	189.30	187.090	181.573	185.936	179.835	188.819
Apr. 2009	175.84	166.394	178.848	180.165	182.514	183.542
MAPE (%)		4.404	3.566	3.799	3.406	1.973

Table 5.14 Wilcoxon signed-rank test

	Wilcoxon signed-rank test	
	$\alpha = 0.025$	$\alpha = 0.05$
Compared models	$W = 2$	$W = 3$
SSVRCCSA vs. SARIMA$(4,2,2)\times(1,2,1)_{12}$	2[a]	2[a]
SSVRCCSA vs. SHW(0.12, 0.95, 0.2, 0.2)	2[a]	2[a]
SSVRCCSA vs. TF-ε-SVR-SA	2[a]	2[a]
SSVRCCSA vs. SVRCCSA	1[a]	1[a]

[a]Denotes that SSVRCCSA model significantly outperforms other alternative models

different forecasting models, including SARIMA$(4,2,2)\times(1,2,1)_{12}$, TF-ε-SVR-SA, SHW(0.12, 0.95, 0.2, 0.2), SVRCSA, and SSVRCSA models, are illustrated in Table 5.13. The proposed SSVRCCSA model with smaller MAPE values is superior to SARIMA$(4,2,2)\times(1,2,1)_{12}$, SHW(0.12, 0.95, 0.2, 0.2), TF-ε-SVR-SA, and SVRCCSA models, due to its capability to excellently learn about the monthly load changing tendency. The seasonal mechanism further revises the forecasting results from the SVRCCSA model (MAPE = 3.406 %), based on the seasonal indexes (per month) obtained from training and validation stages, to achieve more acceptable forecasting accuracy (1.973 %).

For forecasting accuracy improvement significant test, the Wilcoxon signed-rank test and asymptotic test are employed. The test results are shown in Tables 5.14 and 5.15, respectively. Clearly, the SSVRCCSA model receives complete significant forecasting accuracy improvement than SARIMA$(4,2,2)\times(1,2,1)_{12}$, SHW(0.12, 0.95, 0.2, 0.2), TF-ε-SVR-SA, and SVRCCSA models. Particularly for comparing with TF-ε-SVR-SA model (also with seasonal adjustment mechanism but without hybrid evolutionary algorithm and chaotic sequence), the comparison results also recognize that chaotic sequence could significantly improve the performance in terms of premature convergence. By comparing SVRCCSA with SSVRCCSA models, it also indicates the significant superiority from seasonal mechanism. It deserves to pay some attention on those cyclic information while modeling. Figure 5.8 is provided to illustrate the forecasting accuracy among different models.

Table 5.15 Asymptotic test

Compared models	Asymptotic (S_1) test $\alpha = 0.05$	$\alpha = 0.10$
SSVRCCSA vs. SARIMA $(4,2,2) \times (1,2,1)_{12}$	H_0: $e_1 = e_2$ $S_1 = -2.945$; $p = 0.00162$ (reject H_0)	H_0: $e_1 = e_2$ $S_1 = -2.945$; $p = 0.00162$ (reject H_0)
SSVRCCSA vs. SHW(0.12, 0.95, 0.2, 0.2)	H_0: $e_1 = e_2$ $S_1 = -3.066$; $p = 0.00109$ (reject H_0)	H_0: $e_1 = e_2$ $S_1 = -3.066$; $p = 0.00109$ (reject H_0)
SSVRCCSA vs. TF-ε-SVR-SA	H_0: $e_1 = e_2$ $S_1 = -3.788$; $p = 0.00008$ (reject H_0)	H_0: $e_1 = e_2$ $S_1 = -3.788$; $p = 0.00008$ (reject H_0)
SSVRCCSA vs. SVRCCSA	H_0: $e_1 = e_2$ $S_1 = -1.976$; $p = 0.0241$ (reject H_0)	H_0: $e_1 = e_2$ $S_1 = -1.976$; $p = 0.0241$ (reject H_0)

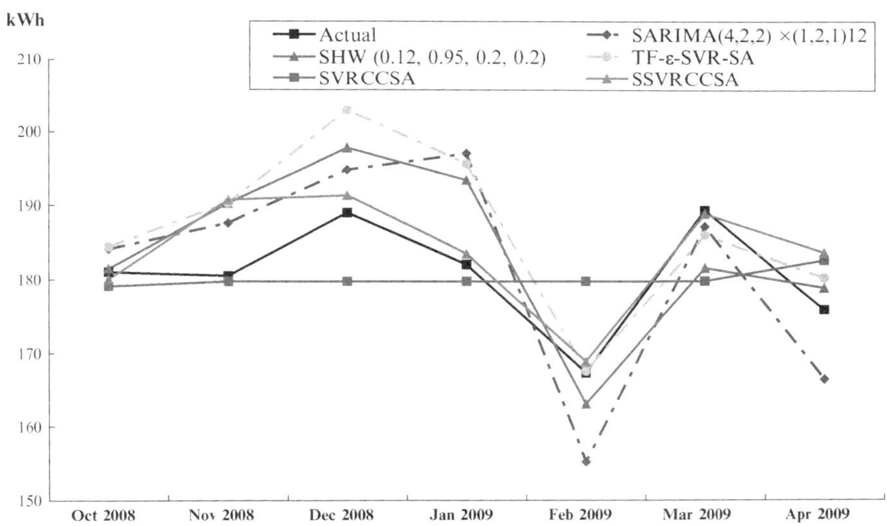

Fig. 5.8 Forecasting results of SARIMA, SHW, TF-ε-SVR-SA, SVRCCSA, and SSVRCCSA models

The significant superiority of the proposed SSVRCCSA model in load forecasting can be summarized as follows. Firstly, the *Y* condition cloud generator can obviously ensure temperature reducing continuously and to overcome the dilemma of the original SA, easily to accept worsened solution, and lead to converge to local minimum while decreasing to low temperature; that is, it can help the original SA to well simulate the actual physical annealing processes, to avoid premature convergence. Secondly, the seasonal mechanism can successfully determine cyclic length and well calculate suitable seasonal indexes for each cycle point.

Table 5.16 The seasonal indexes for each month for the SVRCGASA model

Time point (month)	Seasonal index	Time point (month)	Seasonal index
January	1.0239	July	1.0775
February	0.9180	August	1.0742
March	1.0234	September	1.0189
April	0.9941	October	0.9906
May	1.0271	November	1.0438
June	1.0321	December	1.0694

Table 5.17 Forecasting results of SARIMA, SHW, TF-ε-SVR-SA, SVRCGASA, and SSVRCGASA models (unit: hundred million kWh)

Time point (month)	Actual	SARIMA (4,2,2) × $(1,2,1)_{12}$	SHW(0.12, 0.95, 0.2, 0.2)	TF-ε-SVR-SA	SVRCGASA	SSVRCGASA
Oct. 2008	181.07	184.210	181.554	184.504	177.300	175.639
Nov. 2008	180.56	187.638	190.312	190.361	177.443	185.210
Dec. 2008	189.03	194.915	197.887	202.980	177.585	189.907
Jan. 2009	182.07	197.119	193.511	195.753	177.726	181.970
Feb. 2009	167.35	155.205	163.113	167.580	177.867	163.281
Mar. 2009	189.30	187.090	181.573	185.936	178.008	182.175
Apr. 2009	175.84	166.394	178.848	180.165	178.682	177.629
MAPE (%)		4.404	3.566	3.799	3.731	1.901

5.6 Seasonal Mechanism in SVRCGASA Model and Forecasting Results

The seasonal (cyclic) length of the total employed electric load is also set as 12. Thus, the 12 seasonal indexes are estimated by the 46 in-sample forecasting loads of the SVRCGASA model mentioned in Sect. 4.5.3, including 32 and 14 in-sample forecasting loads in training stage and validation stage, respectively, as shown in Table 5.16. The actual values and the out-of-sample forecasting loads obtained by different forecasting models, including SARIMA(4,2,2) × $(1,2,1)_{12}$, TF-ε-SVR-SA, SHW(0.12, 0.95, 0.2, 0.2), SVRCGASA, and SSVRCGASA models, are illustrated in Table 5.17. The proposed SSVRCGASA model with smaller MAPE values is superior to SARIMA(4,2,2) × $(1,2,1)_{12}$, SHW(0.12, 0.95, 0.2, 0.2), TF-ε-SVR-SA, and SVRCGASA models, due to its capability to excellently learn about the monthly load changing tendency. The seasonal mechanism further revises the forecasting results from the SVRCGASA model (MAPE = 3.731 %), based on the seasonal indexes (per month) obtained from training and validation stages, to achieve more acceptable forecasting accuracy (1.901 %).

For forecasting accuracy improvement significant test, the Wilcoxon signed-rank test and asymptotic test are employed. The test results are shown in Tables 5.18 and 5.19, respectively. Clearly, the SSVRCGASA model receives complete

Table 5.18 Wilcoxon signed-rank test

Compared models	Wilcoxon signed-rank test	
	$\alpha = 0.025$	$\alpha = 0.05$
	$W = 2$	$W = 3$
SSVRCGASA vs. SARIMA(4,2,2) × $(1,2,1)_{12}$	2[a]	2[a]
SSVRCGASA vs.SHW(0.12, 0.95, 0.2, 0.2)	2[a]	2[a]
SSVRCGASA vs. TF-ε-SVR-SA	0[a]	0[a]
SSVRCGASA vs. SVRCGASA	2[a]	2[a]

[a]Denotes that SSVRCGASA model significantly outperforms other alternative models

Table 5.19 Asymptotic test

Compared models	Asymptotic (S_1) test	
	$\alpha = 0.05$	$\alpha = 0.10$
SSVRCGASA vs. SARIMA (4,2,2) × $(1,2,1)_{12}$	H_0: $e_1 = e_2$ $S_1 = -3.329$; $p = 0.000432$ (reject H_0)	H_0: $e_1 = e_2$ $S_1 = -3.329$; $p = 0.000432$ (reject H_0)
SSVRCGASA vs. SHW(0.12, 0.95, 0.2, 0.2)	H_0: $e_1 = e_2$ $S_1 = -17.745$; $p = 0.000$ (reject H_0)	H_0: $e_1 = e_2$ $S_1 = -17.745$; $p = 0.000$ (reject H_0)
SSVRCGASA vs. TF-ε-SVR-SA	H_0: $e_1 = e_2$ $S_1 = -6.222$; $p = 0.000$ (reject H_0)	H_0: $e_1 = e_2$ $S_1 = -6.222$; $p = 0.000$ (reject H_0)
SSVRCGASA vs. SVRCGASA	H_0: $e_1 = e_2$ $S_1 = -2.563$; $p = 0.005185$ (reject H_0)	H_0: $e_1 = e_2$ $S_1 = -2.563$; $p = 0.005185$ (reject H_0)

significant forecasting accuracy improvement than SARIMA(4,2,2) × $(1,2,1)_{12}$, SHW(0.12, 0.95, 0.2, 0.2), TF-ε-SVR-SA, and SVRCGASA models. Particularly for comparing with TF-ε-SVR-SA model (also with seasonal adjustment mechanism but without hybrid evolutionary algorithm and chaotic sequence), the comparison results also recognize that chaotic sequence could significantly improve the performance in terms of premature convergence. By comparing SVRCGASA with SSVRCGASA models, it also indicates the significant superiority from seasonal mechanism, which can successfully determine cyclic length and well calculate suitable seasonal indexes for each cycle point. By the way, it should be noticed that the proposed SSVRCGASA model will implement three processes, including SVR modeling, CGASA conducting, and seasonal mechanism; of course, it will cost some reasonable processing time. However, it deserves to pay some attention on those cyclic information analyses while modeling. Figure 5.9 is provided to illustrate the forecasting accuracy among different models.

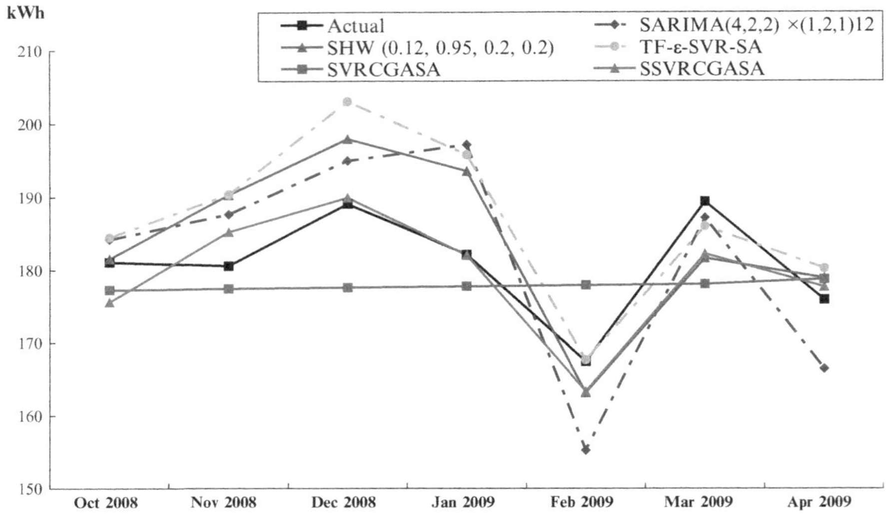

Fig. 5.9 Forecasting results of SARIMA, SHW, TF-ε-SVR-SA, SVRCGASA, and SSVRCGASA models

Table 5.20 The seasonal indexes for each month for the SVRCPSO model

Time point (month)	Seasonal index	Time point (month)	Seasonal index
January	1.0606	July	1.0430
February	1.0170	August	1.0791
March	0.9298	September	1.0784
April	1.0429	October	1.0210
May	1.0046	November	0.9992
June	1.0401	December	1.0545

5.7 Seasonal Mechanism in SVRCPSO Model and Forecasting Results

The seasonal (cyclic) length of the total employed electric load is also set as 12. Thus, the 12 seasonal indexes are estimated by the 46 in-sample forecasting loads of the SVRCPSO model mentioned in Sect. 4.6.3, including 32 and 14 in-sample forecasting loads in training stage and validation stage, respectively, as shown in Table 5.20. The actual values and the out-of-sample forecasting loads obtained by different forecasting models, including SARIMA(4,2,2) × $(1,2,1)_{12}$, TF-ε-SVR-SA, SHW(0.12, 0.95, 0.2, 0.2), SVRCPSO, and SSVRCPSO models, are illustrated in Table 5.21. The proposed SSVRCPSO model with smaller MAPE values is superior to SARIMA(4,2,2) × $(1,2,1)_{12}$, SHW(0.12, 0.95, 0.2, 0.2), TF-ε-SVR-SA, and SVRCPSO models, due to its capability to excellently learn about the monthly load changing tendency. The seasonal mechanism further revises the forecasting

Table 5.21 Forecasting results of SARIMA, SHW, TF-ε-SVR-SA, SVRCPSO, and SSVRCPSO models (unit: hundred million kWh)

Time point (month)	Actual	SARIMA $(4,2,2) \times (1,2,1)_{12}$	SHW(0.12, 0.95, 0.2, 0.2)	TF-ε-SVR-SA	SVRCPSO	SSVRCPSO
Oct. 2008	181.07	184.210	181.554	184.504	181.938	181.7964
Nov. 2008	180.56	187.638	190.312	190.361	182.186	192.1178
Dec. 2008	189.03	194.915	197.887	202.980	182.677	193.742
Jan. 2009	182.07	197.119	193.511	195.753	182.794	185.8846
Feb. 2009	167.35	155.205	163.113	167.580	182.826	169.9838
Mar. 2009	189.30	187.090	181.573	185.936	182.746	190.5905
Apr. 2009	175.84	166.394	178.848	180.165	184.222	185.072
MAPE (%)		4.404	3.566	3.799	3.231	2.699

Table 5.22 Wilcoxon signed-rank test

	Wilcoxon signed-rank test	
	$\alpha = 0.025$	$\alpha = 0.05$
Compared models	$W = 2$	$W = 3$
SSVRCPSO vs. SARIMA$(4,2,2) \times (1,2,1)_{12}$	3	3[a]
SSVRCPSO vs. SHW(0.12, 0.95, 0.2, 0.2)	2[a]	2[a]
SSVRCPSO vs. TF-ε-SVR-SA	3	3[a]
SSVRCPSO vs. SVRCPSO	2[a]	2[a]

[a]Denotes that SSVRCPSO model significantly outperforms other alternative models

results from the SVRCPSO model (MAPE = 3.231 %), based on the seasonal indexes (per month) obtained from training and validation stages, to achieve more acceptable forecasting accuracy (2.699 %).

For forecasting accuracy improvement significant test, the Wilcoxon signed-rank test and asymptotic test are employed. The test results are shown in Tables 5.22 and 5.23, respectively. Clearly, the SSVRCPSO model only receives complete significant forecasting accuracy improvement than SHW(0.12, 0.95, 0.2, 0.2) model, but incomplete significant forecasting accuracy improvement than SARIMA$(4,2,2) \times (1,2,1)_{12}$ and TF-ε-SVR-SA models (only receives significance with $\alpha = 0.05$ level in Wilcoxon test, and all pass with both levels in asymptotic test), and SVRCPSO model (receives significance with both levels in Wilcoxon test, but only receives significance with both $\alpha = 0.10$ level in asymptotic test). Particularly for comparing with TF-ε-SVR-SA model (also with seasonal adjustment mechanism but without hybrid evolutionary algorithm and chaotic sequence), the comparison results also recognize that chaotic sequence could significantly improve the performance in terms of premature convergence. By comparing SVRCPSO with SSVRCPSO models, it also indicates the significant superiority from seasonal mechanism. By the way, it should be noticed that the proposed SSVRCPSO model will implement three processes, including SVR modeling, CPSO conducting, and seasonal mechanism; of course, it will cost some reasonable processing time. However, it deserves to pay some attention on those cyclic information analyses while modeling. Figure 5.10 is provided to illustrate the forecasting accuracy among different models.

Table 5.23 Asymptotic test

Compared models	Asymptotic (S_1) test $\alpha = 0.05$	$\alpha = 0.10$
SSVRCPSO vs. SARIMA (4,2,2) × (1,2,1)$_{12}$	H_0: $e_1 = e_2$ $S_1 = -2.586$; $p = 0.004856$ (reject H_0)	H_0: $e_1 = e_2$ $S_1 = -2.586$; $p = 0.004856$ (reject H_0)
SSVRCPSO vs. SHW(0.12, 0.95, 0.2, 0.2)	H_0: $e_1 = e_2$ $S_1 = -2.177$; $p = 0.01472$ (reject H_0)	H_0: $e_1 = e_2$ $S_1 = -2.177$; $p = 0.01472$ (reject H_0)
SSVRCPSO vs. TF-ε-SVR-SA	H_0: $e_1 = e_2$ $S_1 = -3.266$; $p = 0.000548$ (reject H_0)	H_0: $e_1 = e_2$ $S_1 = -3.266$; $p = 0.000548$ (reject H_0)
SSVRCPSO vs. SVRCPSO	H_0: $e_1 = e_2$ $S_1 = -1.450$; $p = 0.0735$(not reject H_0)	H_0: $e_1 = e_2$ $S_1 = -1.450$; $p = 0.0735$ (reject H_0)

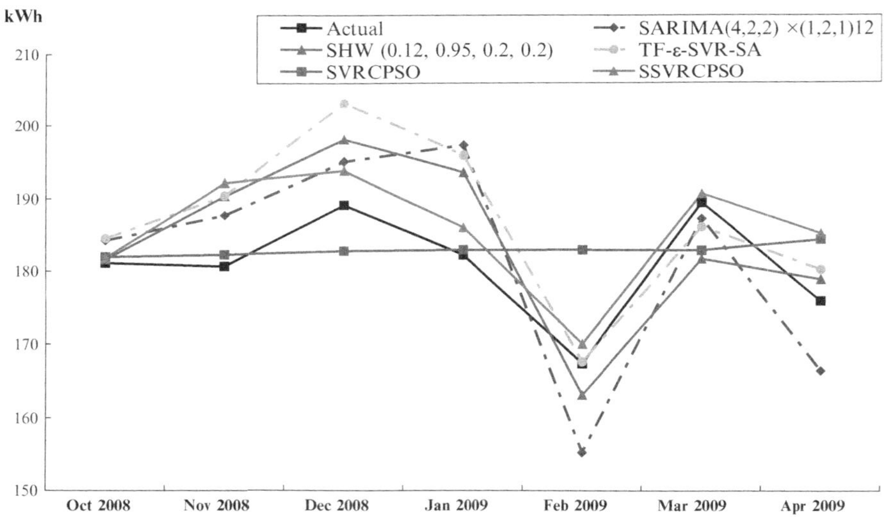

Fig. 5.10 Forecasting results of SARIMA, SHW, TF-ε-SVR-SA, SVRCPSO, and SSVRCPSO models

5.8 Seasonal Mechanism in SVRCAS Model and Forecasting Results

The seasonal (cyclic) length of the total employed electric load is also set as 12. Thus, the 12 seasonal indexes are estimated by the 46 in-sample forecasting loads of the SVRCAS model mentioned in Sect. 4.7.3, including 32 and 14 in-sample forecasting loads in training stage and validation stage, respectively, as shown in Table 5.24. The actual values and the out-of-sample forecasting loads obtained by different forecasting

Table 5.24 The seasonal indexes for each month for the SVRCAS model

Time point (month)	Seasonal index	Time point (month)	Seasonal index
January	1.0311	July	1.0673
February	0.9140	August	1.0617
March	1.0175	September	1.0079
April	0.9906	October	1.0197
May	1.0191	November	1.0362
June	1.0233	December	1.0783

Table 5.25 Forecasting results of SARIMA, SHW, TF-ε-SVR-SA, SVRCAS, and SSVRCAS models (unit: hundred million kWh)

Time point (month)	Actual	SARIMA $(4,2,2) \times (1,2,1)_{12}$	SHW(0.12, 0.95, 0.2, 0.2)	TF-ε-SVR-SA	SVRCAS	SSVRCAS
Oct. 2008	181.07	184.210	181.554	184.504	180.6185	184.1706
Nov. 2008	180.56	187.638	190.312	190.361	180.8985	187.4521
Dec. 2008	189.03	194.915	197.887	202.980	181.1779	195.3663
Jan. 2009	182.07	197.119	193.511	195.753	181.4569	187.0961
Feb. 2009	167.35	155.205	163.113	167.580	181.7354	166.1057
Mar. 2009	189.30	187.090	181.573	185.936	182.0133	185.1910
Apr. 2009	175.84	166.394	178.848	180.165	180.7582	179.0545
MAPE (%)		4.404	3.566	3.799	2.881	2.341

Table 5.26 Wilcoxon signed-rank test

	Wilcoxon signed-rank test	
	$\alpha = 0.025$	$\alpha = 0.05$
Compared models	$W = 2$	$W = 3$
SSVRCAS vs. SARIMA$(4,2,2) \times (1,2,1)_{12}$	3	3[a]
SSVRCAS vs. SHW(0.12, 0.95, 0.2, 0.2)	3	3[a]
SSVRCAS vs. TF-ε-SVR-SA	0[a]	0[a]
SSVRCAS vs. SVRCAS	2[a]	2[a]

[a]Denotes that SSVRCAS model significantly outperforms other alternative models

models, including SARIMA$(4,2,2) \times (1,2,1)_{12}$, TF-ε-SVR-SA, SHW(0.12, 0.95, 0.2, 0.2), SVRCAS, and SSVRCAS models, are illustrated in Table 5.25. The proposed SSVRCAS model with smaller MAPE values is superior to SARIMA $(4,2,2) \times (1,2,1)_{12}$, SHW(0.12, 0.95, 0.2, 0.2), TF-ε-SVR-SA, and SVRCAS models, due to its capability to excellently learn about the monthly load changing tendency. The seasonal mechanism further revises the forecasting results from the SVRCAS model (MAPE = 2.881 %), based on the seasonal indexes (per month) obtained from training and validation stages, to achieve more acceptable forecasting accuracy (2.341 %).

For forecasting accuracy improvement significant test, the Wilcoxon signed-rank test and asymptotic test are also used. The test results are shown in Tables 5.26 and 5.27, respectively. Clearly, the SSVRCAS model only receives complete

Table 5.27 Asymptotic test

	Asymptotic (S_1) test	
Compared models	$\alpha = 0.05$	$\alpha = 0.10$
SSVRCAS vs. SARIMA (4,2,2) × $(1,2,1)_{12}$	H_0: $e_1 = e_2$ $S_1 = -3.477$; $p = 0.000253$ (reject H_0)	H_0: $e_1 = e_2$ $S_1 = -3.477$; $p = 0.000253$ (reject H_0)
SSVRCAS vs. SHW(0.12, 0.95, 0.2, 0.2)	H_0: $e_1 = e_2$ $S_1 = -7.430$; $p = 0.000$ (reject H_0)	H_0: $e_1 = e_2$ $S_1 = -7.430$; $p = 0.000$ (reject H_0)
SSVRCAS vs. TF-ε-SVR-SA	H_0: $e_1 = e_2$ $S_1 = -5.726$; $p = 0.000$ (reject H_0)	H_0: $e_1 = e_2$ $S_1 = -5.726$; $p = 0.000$ (reject H_0)
SSVRCAS vs. SVRCAS	H_0: $e_1 = e_2$ $S_1 = -1.971$; $p = 0.02435$ (reject H_0)	H_0: $e_1 = e_2$ $S_1 = -1.971$; $p = 0.02435$ (reject H_0)

significant forecasting accuracy improvement than TF-ε-SVR-SA and SVRCAS models, but incomplete significant forecasting accuracy improvement than SARIMA(4,2,2) × $(1,2,1)_{12}$ and SHW(0.12, 0.95, 0.2, 0.2) models (only receives significance with $\alpha = 0.05$ level in Wilcoxon test, and all pass with both levels in asymptotic test). Particularly for comparing with TF-ε-SVR-SA model (also with seasonal adjustment mechanism but without hybrid evolutionary algorithm and chaotic sequence), the comparison results also recognize that chaotic sequence could significantly improve the performance in terms of premature convergence. By comparing SVRCAS with SSVRCAS models, it also indicates the significant superiority from seasonal mechanism. By the way, it should be noticed that the proposed SSVRCAS model will implement three processes, including SVR modeling, CAS conducting, and seasonal mechanism; of course, it will cost some reasonable processing time. However, it deserves to pay some attention on those cyclic information analyses while modeling. Figure 5.11 is provided to illustrate the forecasting accuracy among different models.

The proposed SSVRCAS model has obtained significant smaller MAPE values than other alternative models (SARIMA(4,2,2) × $(1,2,1)_{12}$, SHW(0.12, 0.95, 0.2, 0.2), TF-ε-SVR-SA, and SVRCAS models). It is caused by (1) nonlinear mapping capabilities and structural risk minimization of SVR model itself; (2) the CAS algorithm employed the organization variable to perform self-organization foraging process of ant colony to determine proper parameters combination, and applies ergodicity property of chaotic sequences to enrich the searching behavior to avoid premature convergence; (3) the seasonal adjustment with well seasonal/cyclic analytical ability of load demanding tendency.

It is interesting to address the SVRCAS model focuses on the interactions among individual ant's chaotic behavior and ant colony organization foraging activities, instead of "expert rules," to negotiate and to coordinate to look for much better solutions. Therefore, the better solution is evolved with "learning by doing" activities among ants and their colony to approximately reach the global optimum or near-optimum. It is sometimes like a "practical (numerical) rule" to guide colony organization variable that enhances its effects on individual ant's chaotic behaviors.

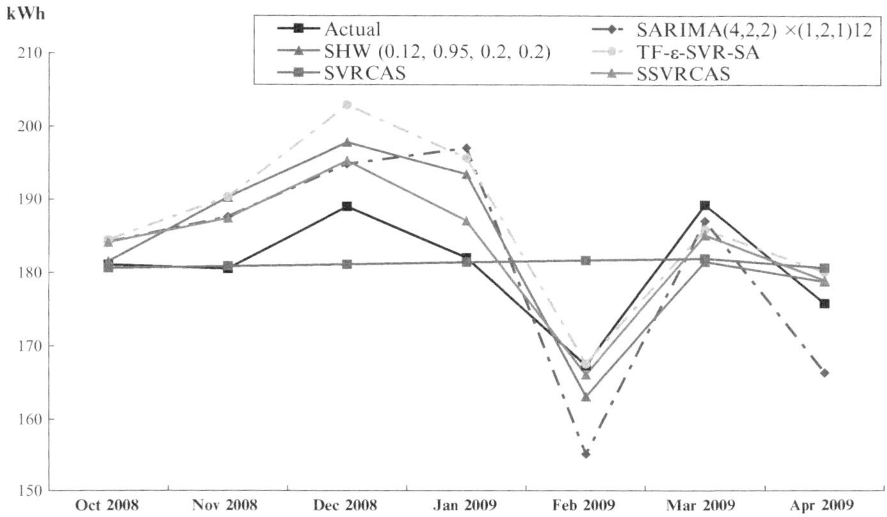

Fig. 5.11 Forecasting results of SARIMA, SHW, TF-ε-SVR-SA, SVRCAS, and SSVRCAS models

Furthermore, for example, along with the unexpected climate change pattern, the tendencies of the electric load data may present more fluctuant nonhistorically, and the future changes of the electric load data may be more cyclic with more short cycle. The proposed SSVRCAS model is potential to approximately reach the near-optimum by "learning by doing" activities among ants and their colony (organization variable), to adjust any length of seasonal load demand (weekly, monthly, bimonthly, quarterly, etc.) by seasonal adjustment; thus, it is much more potential alternative forecasting model in nonhistorical climate change age.

5.9 Seasonal Mechanism in SVRCABC Model and Forecasting Results

The seasonal (cyclic) length of the total employed electric load is also set as 12. Thus, the 12 seasonal indexes are estimated by the 46 in-sample forecasting loads of the SVRCABC model mentioned in Sect. 4.8.3, including 32 and 14 in-sample forecasting loads in training stage and validation stage, respectively, as shown in Table 5.28. The actual values and the out-of-sample forecasting loads obtained by different forecasting models, including SARIMA(4,2,2) $\times$ $(1,2,1)_{12}$, TF-ε-SVR-SA, SHW(0.12, 0.95, 0.2, 0.2), SVRCABC, and SSVRCABC models, are illustrated in Table 5.29. The proposed SSVRCABC model with smaller MAPE values is superior to SARIMA(4,2,2) $\times$ $(1,2,1)_{12}$, SHW(0.12, 0.95, 0.2, 0.2), TF-ε-SVR-SA, and SVRCABC models, due to its capability to excellently learn about

Table 5.28 The seasonal indexes for each month for the SVRCABC model

Time point (month)	Seasonal index	Time point (month)	Seasonal index
January	1.0202	July	1.0875
February	0.9346	August	1.0817
March	1.0448	September	1.0195
April	1.0081	October	1.0045
May	1.0467	November	1.0603
June	1.0467	December	1.0637

Table 5.29 Forecasting results of SARIMA, SHW, TF-ε-SVR-SA, SVRCABC, and SSVRCABC models (unit: hundred million kWh)

Time point (month)	Actual	SARIMA $(4,2,2) \times (1,2,1)_{12}$	SHW(0.12, 0.95, 0.2, 0.2)	TF-ε-SVR-SA	SVRCABC	SSVRCABC
Oct. 2008	181.07	184.210	181.554	184.504	182.131	182.9516
Nov. 2008	180.56	187.638	190.312	190.361	182.788	193.8166
Dec. 2008	189.03	194.915	197.887	202.980	182.791	194.4410
Jan. 2009	182.07	197.119	193.511	195.753	182.793	186.4791
Feb. 2009	167.35	155.205	163.113	167.580	182.795	170.8391
Mar. 2009	189.30	187.090	181.573	185.936	182.747	190.9312
Apr. 2009	175.84	166.394	178.848	180.165	182.772	184.2451
MAPE (%)		4.404	3.566	3.799	3.164	3.056

Table 5.30 Wilcoxon signed-rank test

	Wilcoxon signed-rank test	
	$\alpha = 0.025$	$\alpha = 0.05$
Compared models	$W = 2$	$W = 3$
SSVRCABC vs. SARIMA$(4,2,2) \times (1,2,1)_{12}$	3	3[a]
SSVRCABC vs. SHW(0.12, 0.95, 0.2, 0.2)	2[a]	2[a]
SSVRCABC vs. TF-ε-SVR-SA	3	3[a]
SSVRCABC vs. SVRCABC	1[a]	1[a]

[a]Denotes that SSVRCABC model significantly outperforms other alternative models

the monthly load changing tendency. The seasonal mechanism further revises the forecasting results from the SVRCABC model (MAPE = 3.164 %), based on the seasonal indexes (per month) obtained from training and validation stages, to achieve more acceptable forecasting accuracy (3.056 %).

For forecasting accuracy improvement significant test, the Wilcoxon signed-rank test and asymptotic test are also used. The test results are shown in Tables 5.30 and 5.31, respectively. Clearly, the SSVRCABC model only receives incomplete significant forecasting accuracy improvement than SARIMA$(4,2,2) \times (1,2,1)_{12}$, TF-ε-SVR-SA models (only receives significance with $\alpha = 0.05$ level in Wilcoxon test, and all pass with both levels in asymptotic test), SHW(0.12, 0.95, 0.2, 0.2) and SVRCABC models (receives significance with both levels in Wilcoxon test, and all fails with both levels in asymptotic test). Particularly for comparing with TF-ε-SVR-SA model (also with seasonal adjustment mechanism but without hybrid

Table 5.31 Asymptotic test

	Asymptotic (S_1) test	
Compared models	$\alpha = 0.05$	$\alpha = 0.10$
SSVRCABC vs. SARIMA (4,2,2) × (1,2,1)$_{12}$	H_0: $e_1 = e_2$ $S_1 = -2.075$; $p = 0.019$ (reject H_0)	H_0: $e_1 = e_2$ $S_1 = -2.075$; $p = 0.019$ (reject H_0)
SSVRCABC vs. SHW(0.12, 0.95, 0.2, 0.2)	H_0: $e_1 = e_2$ $S_1 = -1.232$; $p = 0.10894$ (not reject H_0)	H_0: $e_1 = e_2$ $S_1 = -1.232$; $p = 0.10894$ (not reject H_0)
SSVRCABC vs. TF-ε-SVR-SA	H_0: $e_1 = e_2$ $S_1 = -2.446$; $p = 0.00722$ (reject H_0)	H_0: $e_1 = e_2$ $S_1 = -2.446$; $p = 0.00722$ (reject H_0)
SSVRCABC vs. SVRCABC	H_0: $e_1 = e_2$ $S_1 = -0.808$; $p = 0.20958$ (not reject H_0)	H_0: $e_1 = e_2$ $S_1 = -0.808$; $p = 0.20958$ (not reject H_0)

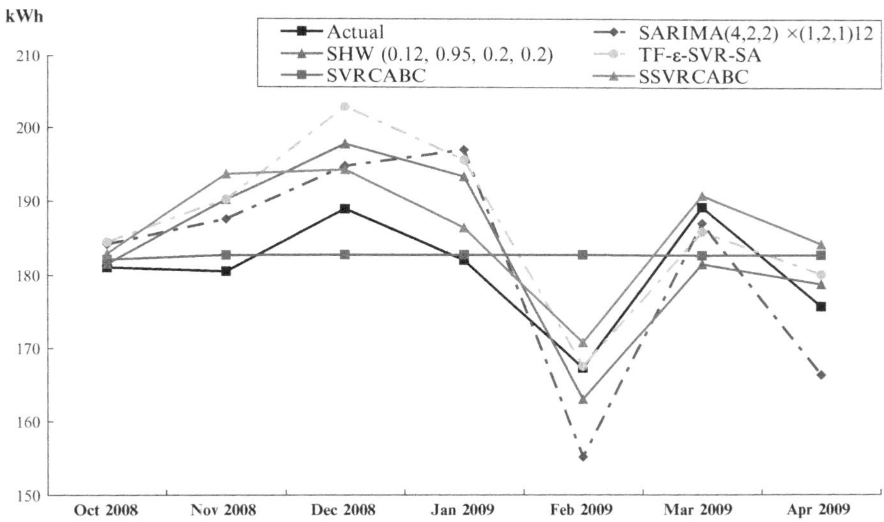

Fig. 5.12 Forecasting results of SARIMA, SHW, TF-ε-SVR-SA, SVRCABC, and SSVRCABC models

evolutionary algorithm and chaotic sequence), the comparison results also recognize that chaotic sequence could significantly improve the performance in terms of premature convergence. By comparing SVRCABC with SSVRCABC models, it also indicates the superiority from seasonal mechanism; it deserves to pay some attention on those cyclic information analyses while modeling. To look for more capability to receive complete significant forecasting accuracy improvement, it is necessary to use the final hybridization tool, recurrent mechanism, as shown in the following subsection. Figure 5.12 is provided to illustrate the forecasting accuracy among different models.

Table 5.32 Parameters determination of RSVRCABC model

	Parameters			
Nos. of fed-in data	σ	C	ε	MAPE of testing (%)
5	107.24	170.69	8.9356	3.232
10	5.89	177.03	2.2860	3.200
15	4.14	9932.70	14.2530	3.016
20	63.00	6326.70	19.1810	3.009
25	50.27	7681.30	19.3750	2.960

The proposed SSVRCABC model has obtained smaller MAPE values than other alternative models (SARIMA(4,2,2) $\times$ $(1,2,1)_{12}$, SHW(0.12, 0.95, 0.2, 0.2), TF-ε-SVR-SA, and SVRCABC models). It is caused by (1) nonlinear mapping capabilities and structural risk minimization of SVR model itself, (2) the CABC algorithm that employs both global search and local search in each iteration to receive better performance and applies ergodicity property of chaotic sequences to enrich the searching behavior to avoid premature convergence, and (3) the seasonal mechanism with well seasonal/cyclic analytical ability of load demanding tendency.

5.10 Recurrent and Seasonal Mechanisms in SVRCABC Model and Forecasting Results

This subsection will firstly demonstrate the three parameters determination of the proposed hybrid model (recurrent SVR with CABC), namely, RSVRCABC model. Secondly, conduct the combined model (RSVRCABC with seasonal mechanism), namely, SRSVRCABC model.

For RSVRCABC modeling process, based on the parameter determination results in Sect. 4.8.3, the SVRCABC model with the smallest testing MAPE value is used further to implement the RSVRCABC model. After performing the RSVRCABC model, the final forecasting values are gained, and the kernel parameters, σ, C, and ε, are used as the most suitable model in this example. The forecasting results and the suitable parameters for the RSVRCABC model are illustrated in Table 5.32, in which it is indicated that these three models all perform the best when 25 fed-in data are used.

Now the seasonal mechanism is considered. The seasonal (cyclic) length of the total employed electric load is also set as 12. Thus, the 12 seasonal indexes are estimated by the 46 in-sample forecasting loads of the SVRCABC model mentioned in Sect. 4.8.3, including 32 and 14 in-sample forecasting loads in training stage and validation stage, respectively, as shown in Table 5.33. The actual values and the out-of-sample forecasting loads obtained by different forecasting models, including SARIMA(4,2,2) $\times$ $(1,2,1)_{12}$, TF-ε-SVR-SA, SHW(0.12, 0.95, 0.2, 0.2), SSVRCABC, RSVRCABC, and SRSVRCABC models, are illustrated in Table 5.34. The proposed SRSVRCABC model with smaller MAPE values is superior to

Table 5.33 The seasonal indexes for each month for the SVRCABC model

Time point (month)	Seasonal index	Time point (month)	Seasonal index
January	1.0336	July	1.0692
February	0.9167	August	1.0648
March	1.0206	September	1.0110
April	0.9923	October	0.9895
May	1.0202	November	1.0415
June	1.0249	December	1.0807

SARIMA(4,2,2) × $(1,2,1)_{12}$, SHW(0.12, 0.95, 0.2, 0.2), TF-ε-SVR-SA, SSVRCABC, and RSVRCABC models, due to its capability to excellently learn about the monthly load changing tendency. The seasonal mechanism further revises the forecasting results from the RSVRCABC model (MAPE = 2.960 %), based on the seasonal indexes (per month) obtained from training and validation stages, to achieve more acceptable forecasting accuracy (2.387 %).

For forecasting accuracy improvement significant test, the Wilcoxon signed-rank test and asymptotic test are also used. The test results are shown in Tables 5.35 and 5.36, respectively. Clearly, the SRSVRCABC model receives complete significant forecasting accuracy improvement than SARIMA(4,2,2) × $(1,2,1)_{12}$, SHW(0.12, 0.95, 0.2, 0.2), TF-ε-SVR-SA, SSVRCABC, and RSVRCABC models. Particularly for comparing with TF-ε-SVR-SA model (also with seasonal adjustment mechanism but without hybrid evolutionary algorithm and chaotic sequence), the comparison results also recognize that chaotic sequence could significantly improve the performance in terms of premature convergence. By comparing RSVRCABC with SRSVRCABC models, it also indicates the superiority from seasonal mechanism; it deserves to pay some attention on those cyclic information analyses while modeling. Figure 5.13 is provided to illustrate the forecasting accuracy among different models.

The proposed SRSVRCABC model has obtained smaller MAPE values than other alternative models (SARIMA(4,2,2) × (1,2,1)12, SHW(0.12, 0.95, 0.2, 0.2), TF-ε-SVR-SA, SSVRCABC, and RSVRCABC models). It is caused by (1) nonlinear mapping capabilities and structural risk minimization of SVR model itself, (2) the CABC algorithm that employs both global search and local search in each iteration to receive better performance and applies ergodicity property of chaotic sequences to enrich the searching behavior to avoid premature convergence, (3) the recurrent mechanism with superior capability to capture more data pattern information from past electric load data, and (4) the seasonal adjustment with well cyclic (seasonal) analytical ability of load demanding tendency. For example, recurrent mechanism, hybridized into the SVRCABC model, also plays a contributive role to further improve the better solution of SVRCABC model to another solution (σ, C, ε) = (50.27, 7681.30, 19.3750) of RSVRCABC model to be the more appropriate optimal solution with forecasting error in terms of MAPE (2.960) (refer to Tables 4.27 and 5.38). Finally, the seasonal mechanism further revises the forecasting results from RSVRCABC model, based on their seasonal indexes (per

Table 5.34 Forecasting results of SARIMA, SHW, TF-ε-SVR-SA, SSVRCABC, RSVRCABC, and SRSVRCABC models (unit: hundred million kWh)

Time point (month)	Actual	SARIMA(4,2,2) × $(1,2,1)_{12}$	SHW(0.12, 0.95, 0.2, 0.2)	TF-ε-SVR-SA	SSVRCABC	RSVRCABC	SRSVRCABC
Oct. 2008	181.07	184.210	181.554	184.504	182.9516	180.315	178.4199
Nov. 2008	180.56	187.638	190.312	190.361	193.8166	180.542	188.0391
Dec. 2008	189.03	194.915	197.887	202.980	194.4410	180.769	195.3528
Jan. 2009	182.07	197.119	193.511	195.753	186.4791	180.995	187.0825
Feb. 2009	167.35	155.205	163.113	167.580	170.8391	181.221	166.1220
Mar. 2009	189.30	187.090	181.573	185.936	190.9312	181.447	185.1950
Apr. 2009	175.84	166.394	178.848	180.165	184.2451	180.926	179.5335
MAPE (%)		4.404	3.566	3.799	3.056	2.960	2.387

Table 5.35 Wilcoxon signed-rank test

	Wilcoxon signed-rank test	
	$\alpha = 0.025$	$\alpha = 0.05$
Compared models	$W = 2$	$W = 3$
SRSVRCABC vs. SARIMA(4,2,2) × $(1,2,1)_{12}$	2[a]	2[a]
SRSVRCABC vs. SHW(0.12, 0.95, 0.2, 0.2)	2[a]	2[a]
SRSVRCABC vs. TF-ε-SVR-SA	0[a]	0[a]
SRSVRCABC vs. SSVRCABC	2[a]	2[a]
SRSVRCABC vs. RSVRCABC	2[a]	2[a]

[a]Denotes that SRSVRCABC model significantly outperforms other alternative models

Table 5.36 Asymptotic test

	Asymptotic (S_1) test	
Compared models	$\alpha = 0.05$	$\alpha = 0.10$
SRSVRCABC vs. SARIMA (4,2,2) × $(1,2,1)_{12}$	H_0: $e_1 = e_2$ $S_1 = -3.417$; $p = 0.000313$ (reject H_0)	H_0: $e_1 = e_2$ $S_1 = -3.417$; $p = 0.000313$ (reject H_0)
SRSVRCABC vs. SHW(0.12, 0.95, 0.2, 0.2)	H_0: $e_1 = e_2$ $S_1 = -5.896$; $p = 0.000$ (reject H_0)	H_0: $e_1 = e_2$ $S_1 = -5.896$; $p = 0.000$ (reject H_0)
SRSVRCABC vs. TF-ε-SVR-SA	H_0: $e_1 = e_2$ $S_1 = -5.355$; $p = 0.000$ (reject H_0)	H_0: $e_1 = e_2$ $S_1 = -5.355$; $p = 0.000$ (reject H_0)
SRSVRCABC vs. SSVRCABC	H_0: $e_1 = e_2$ $S_1 = -1.971$; $p = 0.02435$ (reject H_0)	H_0: $e_1 = e_2$ $S_1 = -1.971$; $p = 0.02435$ (reject H_0)
SRSVRCABC vs. RSVRCABC	H_0: $e_1 = e_2$ $S_1 = -1.960$; $p = 0.025$ (reject H_0)	H_0: $e_1 = e_2$ $S_1 = -1.960$; $p = 0.025$ (reject H_0)

month) obtained from training and validation stages, to achieve more acceptable forecasting accuracies (2.387 %).

It is interesting to address that via recurrent mechanism and seasonal mechanism, the proposed SRSVRCABC model is able to deal with any data pattern no matter data tendencies may present fluctuation or sustained increasing or decreasing types. Furthermore, for example, along with the unexpected climate change pattern, the tendencies of the electric load data or energy-consuming data may present more fluctuant nonhistorically; or along with the large penetration of renewable energies for electricity production, the future changes of the electric load data may be more cyclic with more short cycle. The proposed SRSVRCABC model is potential to approximately reach the global optimum or near-optimum by "communicating in searching" activities among bees and their colony, to learn more fluctuant changed load demand by recurrent mechanism, to adjust any length of seasonal load demand (weekly, monthly, bimonthly, quarterly, etc.) by seasonal mechanism; thus, it is much more potential alternative forecasting model in nonhistorical climate change age.

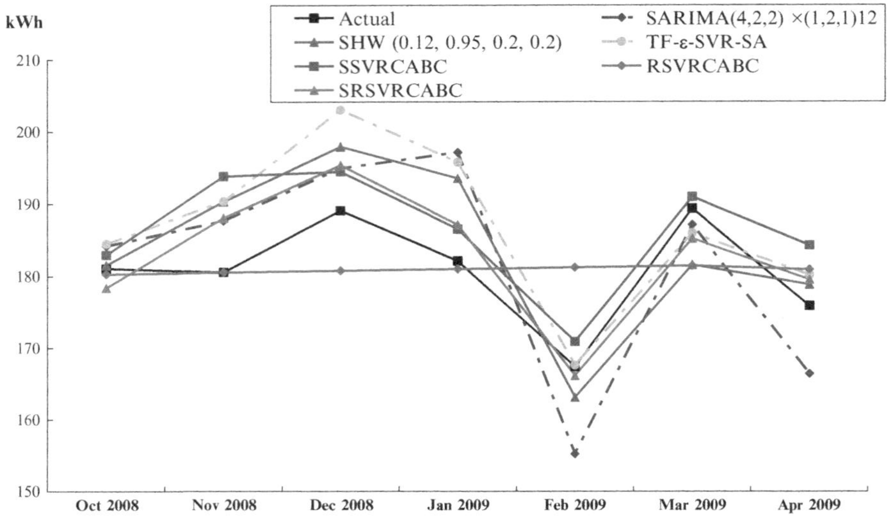

Fig. 5.13 Forecasting results of SARIMA, SHW, TF-ε-SVR-SA, SSVRCABC, RSVRCABC, and SRSVRCABC models

Table 5.37 The seasonal indexes for each month for the SVRCIA model

Time point (month)	Seasonal index	Time point (month)	Seasonal index
January	1.0153	July	1.0663
February	0.9089	August	1.0615
March	1.0126	September	1.0076
April	0.9853	October	0.9734
May	1.0187	November	1.0247
June	1.0225	December	1.0614

5.11 Seasonal Mechanism in SVRCIA Model and Forecasting Results

The seasonal (cyclic) length of the total employed electric load is also set as 12. Thus, the 12 seasonal indexes are estimated by the 46 in-sample forecasting loads of the SVRCIA model mentioned in Sect. 4.9.3, including 32 and 14 in-sample forecasting loads in training stage and validation stage, respectively, as shown in Table 5.37. The actual values and the out-of-sample forecasting loads obtained by different forecasting models, including SARIMA(4,2,2) $\times$ $(1,2,1)_{12}$, TF-ε-SVR-SA, SHW(0.12, 0.95, 0.2, 0.2), SVRCIA, and SSVRCIA models, are illustrated in Table 5.38. The proposed SSVRCIA model with smaller MAPE values is superior to SARIMA(4,2,2) $\times$ $(1,2,1)_{12}$, SHW(0.12, 0.95, 0.2, 0.2), TF-ε-SVR-SA, and SVRCIA models, due to its capability to excellently learn about the monthly load changing tendency. The seasonal mechanism further revises the forecasting results from the SVRCIA model (MAPE = 3.041 %), based on the seasonal indexes (per

Table 5.38 Forecasting results of SARIMA, SHW, TF-ε-SVR-SA, SVRCIA, and SSVRCIA models (unit: hundred million kWh)

Time point (month)	Actual	SARIMA (4,2,2) × $(1,2,1)_{12}$	SHW(0.12, 0.95, 0.2, 0.2)	TF-ε-SVR-SA	SVRCIA	SSVRCIA
Oct. 2008	181.07	184.210	181.554	184.504	179.028	174.274
Nov. 2008	180.56	187.638	190.312	190.361	179.412	183.844
Dec. 2008	189.03	194.915	197.887	202.980	179.795	190.837
Jan. 2009	182.07	197.119	193.511	195.753	180.176	182.934
Feb. 2009	167.35	155.205	163.113	167.580	180.556	164.106
Mar. 2009	189.30	187.090	181.573	185.936	180.934	183.211
Apr. 2009	175.84	166.394	178.848	180.165	178.104	175.483
MAPE (%)		4.404	3.566	3.799	3.041	1.766

Table 5.39 Wilcoxon signed-rank test

	Wilcoxon signed-rank test	
	$\alpha = 0.025$	$\alpha = 0.05$
Compared models	$W = 2$	$W = 3$
SSVRCIA vs. SARIMA(4,2,2) × $(1,2,1)_{12}$	2[a]	2[a]
SSVRCIA vs. SHW(0.12, 0.95, 0.2, 0.2)	2[a]	2[a]
SSVRCIA vs. TF-ε-SVR-SA	0[a]	0[a]
SSVRCIA vs. SVRCIA	3	3[a]

[a]Denotes that SSVRCIA model significantly outperforms other alternative models

Table 5.40 Asymptotic test

	Asymptotic (S_1) test	
Compared models	$\alpha = 0.05$	$\alpha = 0.10$
SSVRCIA vs. SARIMA (4,2,2) × $(1,2,1)_{12}$	H_0: $e_1 = e_2$ $S_1 = -3.091$; $p = 0.00097$ (reject H_0)	H_0: $e_1 = e_2$ $S_1 = -3.091$; $p = 0.00097$ (reject H_0)
SSVRCIA vs. SHW(0.12, 0.95, 0.2, 0.2)	H_0: $e_1 = e_2$ $S_1 = -20.751$; $p = 0.000$ (reject H_0)	H_0: $e_1 = e_2$ $S_1 = -20.751$; $p = 0.000$ (reject H_0)
SSVRCIA vs. TF-ε-SVR-SA	H_0: $e_1 = e_2$ $S_1 = -5.692$; $p = 0.000$ (reject H_0)	H_0: $e_1 = e_2$ $S_1 = -5.692$; $p = 0.000$ (reject H_0)
SSVRCIA vs. SVRCIA	H_0: $e_1 = e_2$ $S_1 = -1.797$; $p = 0.03614$ (reject H_0)	H_0: $e_1\ n\ e_2$ $S_1 = -1.797$; $p = 0.03614$ (reject H_0)

month) obtained from training and validation stages, to achieve more acceptable forecasting accuracy (1.766 %).

For forecasting accuracy improvement significant test, the Wilcoxon signed-rank test and asymptotic test are also used. The test results are shown in Tables 5.39 and 5.40, respectively. Clearly, the SSVRCIA model almost receives complete

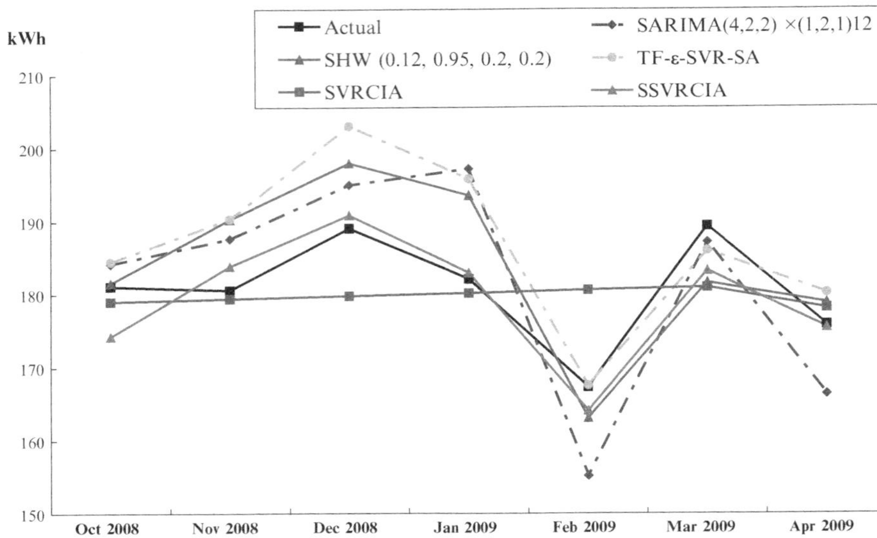

Fig. 5.14 Forecasting results of SARIMA, SHW, TF-ε-SVR-SA, SVRCIA, and SSVRCIA models

significant forecasting accuracy improvement than other alternative models except the SVRCIA model (only receives significance with $\alpha = 0.05$ level in Wilcoxon test and pass both levels in asymptotic test). Particularly for comparing with TF-ε-SVR-SA model (also with seasonal adjustment mechanism but without hybrid evolutionary algorithm and chaotic sequence), the comparison results also recognize that chaotic sequence could significantly improve the performance in terms of premature convergence due to the superior searching capability of CIA to determine proper parameters in an SVR model and the use of a seasonal mechanism to adjust the seasonal/cyclic effects of electric loads. By comparing SVRCIA with SSVRCIA models, it also indicates the superiority from seasonal mechanism employed here is proficient in dealing with such cyclic data types; thus, it deserves to pay some attention on those cyclic information analyses while modeling. Figure 5.14 is provided to illustrate the forecasting accuracy among different models.

The proposed SSVRCIA model has obtained smaller MAPE values than other alternative models (SARIMA(4,2,2) $\times$ $(1,2,1)_{12}$, SHW(0.12, 0.95, 0.2, 0.2), TF-ε-SVR-SA, and SVRCIA models). The superior performance of the SSVRCIA model is not only because of its theoretical assumptions of a convex set while SVR modeling but also because of the superior searching capability of CIA to determine the proper parameters in SVR (this is why it outperforms the TF-ε-SVR-SA model) and effective seasonal mechanism (this is why it outperforms the SVRCIA model). By contrast, SARIMA model employs the parametric technique which is based on specific assumptions, such as linear relationships between the current value of the underlying variables and previous values of the variable and error terms, and these assumptions are not completely in line with real-world problems.

References

1. Kechriotis G, Zervas E, Manolakos ES (1994) Using recurrent neural networks for adaptive communication channel equalization. IEEE Trans Neural Netw 5:267–278. doi:10.1109/72.279190
2. Jordan MI (1986) Attractor dynamics and parallelism in a connectionist sequential machine. In: Proceedings of the 8th annual conference of the cognitive science society, New Jersey, NJ, pp 531–546
3. Elman JL (1990) Finding structure in time. Cogn Sci 14:179–211. doi:10.1207/s15516709cog1402_1
4. Williams R, Zipser D (1989) A learning algorithm for continually running fully recurrent neural networks. Neural Comput 1:270–280. doi:10.1162/neco. 1989.1.2.270
5. Tsoi AC, Back AD (1994) Locally recurrent globally feedforward networks: a critical review of architectures. IEEE Trans Neural Netw 5:229–239. doi:10. 1109/72.279187
6. Jhee WC, Lee JK (1993) Performance of neural networks in managerial forecasting. Int J Intell Syst Acc Financ Manag 2:55–71
7. Suykens JAK, van Gestel T, De Brabanter J, De Moor B, Vandewalle J, Leu-ven KU (2002) Least squares support vector machines. World Scientific Publishing, Belgium
8. Connor JT, Martin RD, Atlas LE (1994) Recurrent neural networks and robust time series prediction. IEEE Trans Neural Netw 5:240–254. doi:10.1109/72. 279188
9. Gencay R, Liu T (1997) Nonlinear modeling and prediction with feedforward and recurrent networks. Physica D 108:119–134. doi:10.1016/S0167-2789(97) 82009-X
10. Kermanshahi B (1998) Recurrent neural network for forecasting next 10 years loads of nine japanese utilities. Neurocomputing 23:125–133. doi:10.1016/S0925-2312(98)00073-3
11. Mandic DP, Chambers JA (2001) Recurrent neural networks for prediction. Wiley, New York, NY
12. Martens K, Chang YC, Taylor S (2002) A comparison of seasonal adjustment methods when forecasting intraday volatility. J Financ Res 25:283–299. doi:10. 1111/1475-6803.t01-1-00009
13. Taylor SJ, Xu X (1997) The incremental volatility information in one million foreign exchange quotations. J Empir Financ 4:317–340. doi:10.1016/S0927- 5398(97)00010-8
14. Andersen TG, Bollerslev T (1998) DM-dollar volatility: intraday activity patterns, macroeconomic announcements and longer run dependencies. J Financ 53:219–265. doi:10.1111/0022-1082.85732
15. Deo R, Hurvich C, Lu Y (2006) Forecasting realized volatility using a long- memory stochastic volatility model: estimation, prediction and seasonal adjustment. J Econom 131:29–58. doi:10.1016/j.jeconom.2005.01.003
16. Wang J, Zhu W, Zhang W, Sun D (2009) A trend fixed on firstly and seasonal adjustment model combined with the ε-SVR for short-term forecasting of electricity demand. Energ Policy 37:4901–4909. doi:10.1016/j.enpol.2009.06.046

CPSIA information can be obtained at www.ICGtesting.com
Printed in the USA
LVOW010258280313

326423LV00006B/70/P

9 781447 149675